Process flowsheeting

Process flowsheeting

A. W. WESTERBERG
PROFESSOR OF CHEMICAL ENGINEERING, CARNEGIE-MELLON
UNIVERSITY

H. P. HUTCHISON
LECTURER IN CHEMICAL ENGINEERING, UNIVERSITY OF
CAMBRIDGE

R. L. MOTARD
PROFESSOR OF CHEMICAL ENGINEERING, WASHINGTON
UNIVERSITY IN ST LOUIS

P. WINTER
MANAGER, PROCESS INDUSTRIES GROUP, COMPUTER AIDED
DESIGN CENTRE, CAMBRIDGE

CAMBRIDGE UNIVERSITY PRESS

CAMBRIDGE

LONDON · NEW YORK · MELBOURNE

CAMBRIDGE UNIVERSITY PRESS
Cambridge, New York, Melbourne, Madrid, Cape Town,
Singapore, São Paulo, Delhi, Tokyo, Mexico City

Cambridge University Press
The Edinburgh Building, Cambridge CB2 8RU, UK

Published in the United States of America by Cambridge University Press, New York

www.cambridge.org
Information on this title: www.cambridge.org/9780521279154

First published 1979
First paperback edition 2011

A catalogue record for this publication is available from the British Library

Library of Congress Cataloguing in Publication data
Main entry under title:
Process flowsheeting.
Bibliography: p.
Includes index.
1. Chemical process control.   2. Chemical plants –
Automation.   1. Westerberg, A. W.
TP155.75.P75   660.2'81   78-56182

ISBN 978-0-521-22043-9 Hardback
ISBN 978-0-521-27915-4 Paperback

Contents

Preface   vii

1   Introduction   1
1.1 Steady-state flowsheeting and the design process   1
1.2 The total design project   5

2   Flowsheeting on the computer   9
2.1 Motivation for development   10
2.2 Developing a simulation model   12
2.3 Approaches to flowsheeting systems – examples   13

3   Solving linear and nonlinear algebraic equations   27
3.1 Solving one equation in one unknown   27
3.2 Solution methods for linear equations   35
3.3 General approaches to solving sets of nonlinear equations   48
3.4 Solving sets of sparse nonlinear equations   77

4   Physical property service facilities   102
4.1 The data cycle   102
4.2 Computerized physical property systems   105
4.3 Physical property calculations   109

5   Degrees of freedom in a flowsheet   113
5.1 Degrees of freedom   113
5.2 Independent stream variables   115
5.3 Degrees of freedom for a unit   120
5.4 Degrees of freedom in a flowsheet   127

6   The sequential modular approach to flowsheeting   130
6.1 The solution of an example flowsheeting problem   131

6.2 Other features   138
6.3 Convergence of tear streams   143
6.4 Partitioning and tearing a flowsheet   148

7   Flowsheeting by equation-solving methods based on tearing   161
7.1 A simple example   162
7.2 An example system based on equation solving   165
7.3 A complex example of selecting decision and tear variables
     for a flowsheet   180
7.4 Handling the iterated variables   189
7.5 Discussion   191

8   Simulation by linear methods   194
8.1 Introduction to linear simulation   194
8.2 Application to staged operations   195
8.3 Application to a management problem   200
8.4 The SYMBOL system for material balancing in linear systems   207
8.5 A simple example of the use of SYMBOL   209
8.6 A complex example of the use of SYMBOL   213

9   Simulation by quasi-linear methods   218
9.1 Introduction to quasi-linear methods   218
9.2 Simulation of flows in pipe networks   220
9.3 Application to distillation   225
9.4 Application to multiple reaction equilibrium   232
9.5 Towards process simulation by quasi-linear methods   238

10   Further reading and literature references   240

     Index   245

Preface

The Computer Aided Design Centre in Cambridge conceived the idea for this book in 1973 and the groundwork was completed in 1974–5 when Professor Art Westerberg and Professor Rudy Motard spent sabbatical appointments at the Centre. Originally, the aim was to publish a short monograph summarizing the status of flowsheeting technology, but it soon became evident that the subject demanded a more considered treatment. Consequently, much additional effort has been devoted to the writing and Art Westerberg in particular applied his enormous energy to revising the text. *Process Flowsheeting* is the result; we hope it is as stimulating in reading as it has been in writing.

The material in this book describes flowsheeting in chemical engineering. As used here the term 'flowsheeting' is the use of computer aids to perform steady-state heat and material balancing, and sizing and costing calculations for a chemical process. Most previous presentations of similar material have concentrated on a single approach to flowsheeting when in fact many exist and each has its distinct advantages and disadvantages. This book presents four approaches, and for each one the motivation for its development is analysed and its use is illustrated through a number of practical examples.

This book can be used to introduce the ideas behind flowsheeting to the novice; it also contains material of interest to anyone who has previously used a flowsheeting system. Lastly some material is aimed directly at those persons who may be thinking about designing a new system or have a burning desire to understand how such systems operate.

The book is organized into four basic parts. The first two chapters are introductory, chapter 1 defining the role of flowsheeting in process design and chapter 2 presenting an introduction to the various approaches that are used in flowsheeting systems to set up and perform the design calculations.

Chapters 3 to 5 give useful background information common to all approaches. Chapter 3 is extensive and deals with solving the large number of simultaneous linear and nonlinear sparse algebraic equations arising in this technology. Chapter 4 draws attention to the calculation of physical property data in process design. The treatment here is introductory since this important and complex area deserves attention in its own right. A companion book which will cover this topic in depth, specifically from the viewpoint of computer-aided design, is planned. Chapter 5 introduces the analysis of degrees of freedom in a flowsheet, a problem that requires understanding to appreciate fully the conceptual problems of designing a flowsheeting system.

The next four chapters, chapters 6 to 9, deal in some detail with each of four approaches used by different flowsheeting systems. Chapter 6 describes the most common approach, the *sequential modular approach*, which underlies such systems as PACER (Digital Systems, 1971), CHESS (Motard and Lee, 1971), CONCEPT (CADC, 1973*a*), and FLOWTRAN (Seader, Seider and Pauls, 1974), as well as many, many others (see Flower and Whitehead, 1973*a*, *b*). Chapter 7 describes flowsheeting by equation-solving methods based on tearing. Few systems exist which use this approach; SPEED-UP (Leigh, Jackson and Sargent, 1974) is perhaps the most widely known. Chapter 8 describes simulation by linear methods, whereby the entire flowsheet is described only approximately using linear equations. This approach permits rapid assessment of process alternatives where detailed analysis is not required. The SYMBOL system (CADC, 1973*b*) illustrates this approach. Chapter 9 describes solving flowsheeting problems using quasi-linear methods such as Newton–Raphson-based methods, and reference is made to the MULTICOL system (CADC, 1977).

The last section of the book, chapter 10, is a guide to further reading material as well as a list of references used throughout the book.

As stated before, this book can be used at several levels. It can be used to introduce flowsheeting to senior students in a design course, when chapters 1, 2, 4 and 6 will be of value. With this background the students will appreciate the role of flowsheeting in process design and will understand why alternative approaches exist. In addition they will have studied the sequential modular approach in some detail. The more advanced material from section 6.2.3. onward may best be omitted from an introductory course of this type.

For those persons with a more advanced background in flowsheeting, chapter 1 may be omitted or at most skimmed through quickly. Chapter 2, however, represents essential reading as it motivates the organization of the book.

Depending on the personal interests of the reader the rest of the book may be approached either by reading chapters 3 to 5 or by proceeding directly to any of chapters 6, 7 or 8, each of which has been written as a self-contained unit. It should be possible to understand all of chapter 6 without first reading chapters 3 to 5. However, the reader should omit section 7.3 if chapter 3 and perhaps chapter 5 have not been studied; the remainder should offer little difficulty. Chapter 8 can also be read and most of the important points understood without the material in chapters 3 to 5.

Chapter 9 is best read only after reading the material contained in chapter 3 on the solving of sets of linear and nonlinear equations.

The nature of the book is to introduce the concepts involved, largely by example. Many of the ideas are therefore not developed here to the same extent as in some existing literature; however, the book is intended to cover more than is currently readily available. It is hoped that the motivations for the various approaches to flowsheeting in existence and for those being developed will be made clear to the reader so he can more fully appreciate the strengths and weaknesses of the plethora of alternatives he will most certainly face in the near future.

October 1977   P. Winter

1

Introduction

1.1 Steady-state flowsheeting and the design process

Flowsheeting may be defined as the use of computer aids to perform steady-state heat and mass balancing, sizing and costing calculations for a chemical process. We shall now detail our definition of this term by investigating for which steps it is used within the design process. Figure 1.1 illustrates the partitioning of the design step. This diagram is in fact useful for design in any field if the word 'flowsheet' is replaced by 'structure', and it is useful for more than viewing steady-state design. However, here we shall restrict our viewpoint to steady-state chemical-process design; later in this chapter we shall discuss a more comprehensive view (e.g. to consider dynamic behaviour) but shall not pursue it beyond that discussion.

Figure 1.1 partitions the process design effort into three basic steps: synthesis, analysis and optimization. Synthesis is the step where the flowsheet structure is chosen, i.e. the particular equipment to be used and its interconnection are selected. Also in the synthesis step one provides initial values for the variables which one is free to set. The second step is 'analysis' and it is often broken into three parts: solving the heat and material balances, sizing and costing the equipment, and evaluating the worth (and perhaps safety, operability and so forth) of the flowsheet just analysed.

The final steps are termed optimization and, in particular, parameter and structural optimization. In the course of analysing a given flowsheet, one usually discovers some particular pressure or temperature level can profoundly influence the resulting equipment sizes and thus the flowsheet evaluation. Or, one may decide to alter the equipment and/or its interconnection because an improvement is obvious or the current version appears to be very costly. Changing the equipment type and/or its interconnection is structural optimization while simply altering temperature or pressure levels within a fixed flowsheet is parameter optimization. The theory to do the latter automatically is in a more advanced state than the former.

The final flowsheet with its decision variable values and resulting flows etc. represents the final design.

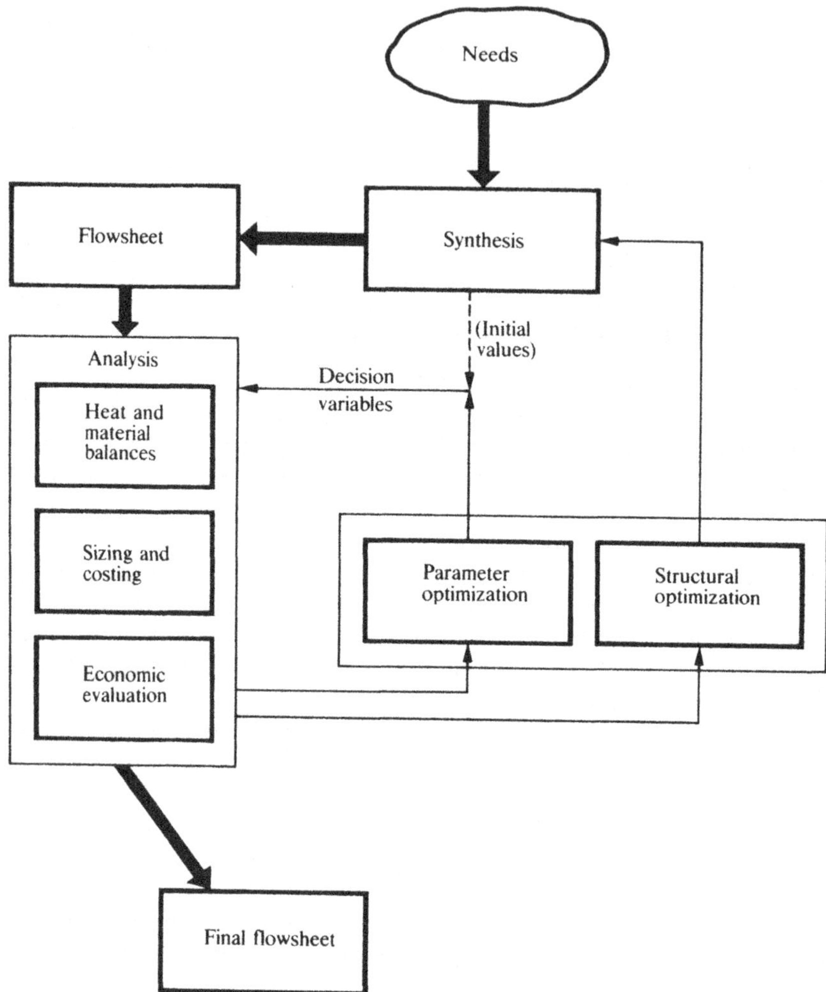

Figure 1.1. Partitioning the process design task into interrelated subtasks.

For this book then, the material to be covered is that describing the computer aids needed only for the analysis step: heat and material balancing, sizing and costing, and preliminary evaluation. The effect of the approaches possible for analysis on the synthesis and optimization steps will be indicated where appropriate, but these last two steps are not within the intended scope of the book.

At this point it might be useful to describe the use of flowsheeting programs within the context of doing a process design. The scenario following is only one of many ways to use such systems.

Scenario

In the initial stages of process development, the engineering team may wish to perform some material balance calculations on a crude flowsheet. The flowsheet will likely contain reactors which do not operate at 100% conversion so the flowsheet will contain recycles, purges and so forth, and in this analysis the approximate flows corresponding to different conversion levels may be desired. Also one may wish to investigate the required purge rates to see what loss of product may be resulting by their use rather than by using more sophisticated separation equipment.

With a proper problem formulation, one can often describe the process adequately for this type of analysis by a set of linear equations. Solving them quickly and repeatedly for slight changes to the structure and conversion levels allows the engineering team to get a feel for the flowsheet, even if the feel is more qualitative than quantitative. The setting up and solving of these linear equations by hand is tedious and error prone, and often impossible because the problem is too large. Also, being certain not to put too many or too few requirements on the flowsheet is difficult. It is here that one can visualize the usefulness of a computer programming system to aid in inputting the flowsheet, the needed data, the required flows and so forth and then to solve, giving diagnostics if a failure occurs. The computer is not doing something here that the engineer could not do in theory, but it is certainly reducing the tedium associated with the job.

At this level of development the engineering team may discover the need to do more actual laboratory development to improve the flowsheet. Thus this aid is used to permit gross effects to be caught early and accounted for, if possible.

Management may now authorize a more detailed flowsheet costing to be performed. The design team will add more detail to the flowsheet, converting each simple separator block into a subsystem of actual separators which will hopefully operate as needed. Now more accurate physical property data will be needed as well as the handling of the heat balances. Simple heaters and coolers, and simple pressure changers will manipulate temperature and pressures as needed, giving the engineers an idea of the heat and power requirements within the process.

The use of a more sophisticated flowsheeting system is required, one which has the ability to calculate physical properties of mixtures and which contains the programs to solve the equations resulting from the types of equipment in the proposed flowsheet.

A considerable effort will be needed to set up the problem, particularly if some of the chemical-component physical property data are not available within the system and if some of the equipment types are new.

The latter will require programs to be written, tested and then included in the system by the engineering team. If the problem is large enough and the best physical property estimation available is desired, then this effort will very probably be less than that for hand calculations. If flowsheet variations are to be tried, the effort will almost certainly be less.

The final flowsheeting calculations to be tried might be those using the most sophisticated design programs available to the system, together with the best procedures for estimating physical properties. Also the flowsheet may be heat and pressure integrated, giving rise to process interactions which are reflected in computational interactions. The costs for such a calculation may well be two-to-three orders of magnitude greater than the initial linear material balance calculations. By this time, however, one is fairly fixed on the flowsheet and is verifying that it operates as needed, to the extent that computer programs can accurately model the process behaviour.

The advantages for performing flowsheeting calculations on the computer do not always include a saving in manpower and design costs. Rather the real advantages may be that the results are much more consistent because the physical property estimation procedures and data are consistent throughout the flowsheet. Parameter variations are more likely to give meaningful trends because of this consistency when doing multiple runs. Also the chances for simple errors in arithmetic are significantly reduced, giving one a better opportunity to eliminate such errors. Once a program has been made as error free as possible, it will blindly, but correctly, perform the calculations it represents from then on.

Comparisons of alternative designs by different design teams are more likely to be realistic because, again, the same design equations and physical property data can be used for both.

The disadvantages are not to be overlooked. The system at hand may be ill-suited for the problem under consideration and may never reach an answer for it, or do so only after very considerable effort. A skilled engineer may, however, be able to use adequate shortcut design methods by hand to get an answer by knowing intuitively how to solve the problem. Or the design team may be tempted to be too ambitious with this exciting new tool available and may never get at least one version of the problem solved by the scheduled time. Flowsheeting systems are man-made and as such contain 'bugs' (errors in the program) which the user may be unaware of, or all-too-aware of, and these 'bugs' may lead to incorrect answers or no answers. Having been 'burnt' once, the user may well be unwilling to try again. As with all large-scale programming systems, the system is likely to be useful only if maintained and if the user has access to expert help if problems do occur.

Following, or along with, the steady-state flowsheeting effort, the dynamic characteristics have to be investigated to establish controller locations and types, and to establish operability during start-up and in the face of expected upsets. Controllers are usually designed by an analysis of small fluctuations in the process, often using a linear dynamic analysis. Operability studies are usually concerned with large process fluctuations, and these studies are needed to locate and size spares, process hold-ups and the like which cannot be established by a steady-state analysis.

1.2 The total design project

Flowsheeting, as we have just described it, is seen to occur during the early stages of the total design project. Figure 1.2 is an illustration giving the steps which occur in a typical project, starting with the conception of the project and ending with the construction of the resulting plant. Throughout this entire process many instances occur when appropriate aids on the computer do, or will in the future, exist.

Looking at figure 1.2, the project starts when the process is in conceptual form, i.e. the need and an approach to meeting it are stated. At this point the *process design* steps start and may be viewed as we have already discussed in the previous section. The end result of this effort is the *process flow diagram* (PFD). A PFD contains the process units and major control elements in a 'functional' form (listed functionally as pumps, columns, reactors, etc.) and the connecting streams. As we have seen, this stage will have the process heat, material and pressure balances solved and will have all units roughly sized and roughly costed.

At this stage the effort moves into the step termed *project engineering*, where the functional description for the process on the PFD is turned into a list of actual equipment to be purchased or constructed, and later enables a set of blueprints or equivalent to be prepared for the actual construction. During project engineering the PFD is first converted into a set of *piping and instrumentation diagrams* (PIDs), often also termed *engineering line diagrams* (ELDs).

The PIDs are a graphical summary of the actual hardware elements in a chemical process plant and their interrelationships, i.e. how they are connected to form an operable, safe and reliable plant. The PIDs, as the hardware reality of the PFD, include vessels (columns and tanks), pipes, valves, pumps, heat exchangers, reactors, furnaces, compressors, expanders, relief and drain valves, traps, filters, conveyors, hoppers, purchased subsystems, sensors, insulation requirements, controllers (flow, pressure, temperature, level), spares and other manufactured items, all in a logical configuration. The PIDs also include vessel sizes,

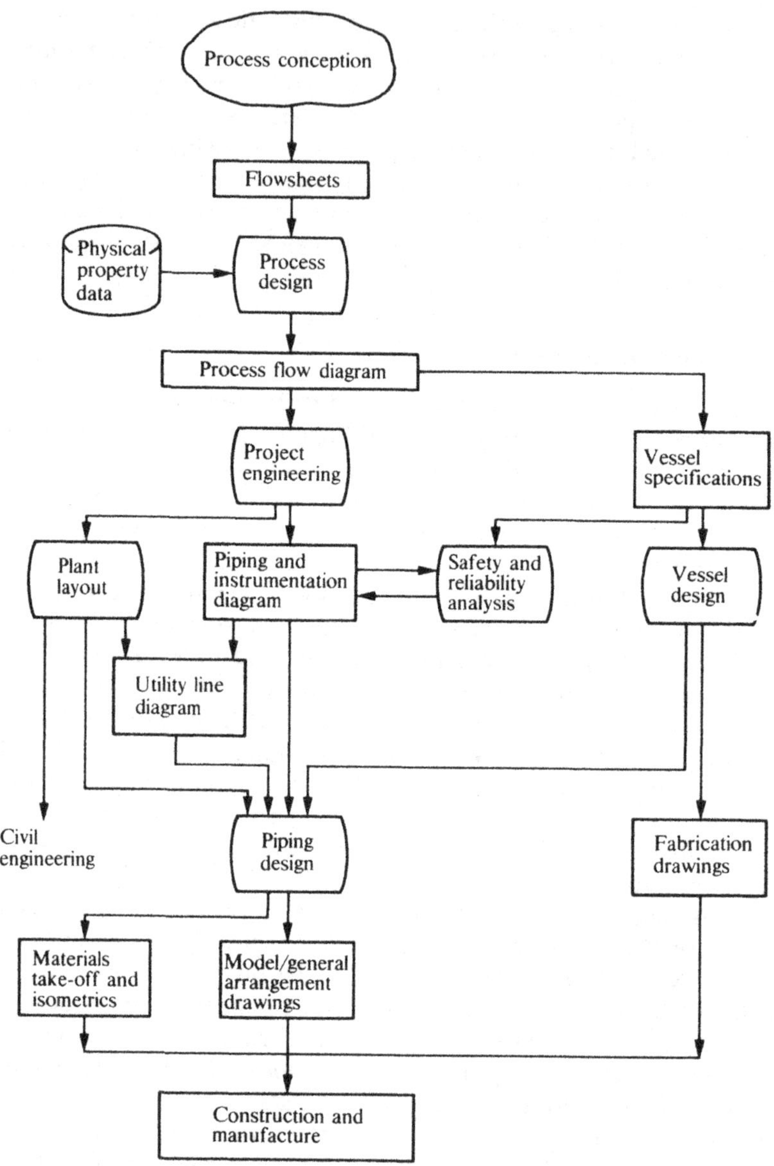

Figure 1.2. The total plant design project.

volumes, pipe diameters, materials of construction, temperatures, pressures and possibly weights of elements. Piping lengths and bends etc. are not included.

While the impression may be conveyed that the PIDs are developed in sequence with other project engineering activities, in fact, much of the project work proceeds in parallel. For instance, with reference to figure 1.2, to obtain pipe diameters an estimate of length is needed so that pressure drop may be calculated. This implies that a preliminary plant layout has been done which in turn depends on the routing of one or more critical pipes (high temperature, vapour or two-phase transfer lines). These in turn have been obtained through a preliminary pipe design analysis. It also happens frequently that the PIDs cannot be, or are not, completed until project engineering is well advanced.

On figure 1.2 an additional diagram type similar to the PID is indicated; it is called the *utility line diagram* (ULD), and includes hardware details of the steam and water piping and control systems. These diagrams are even more sensitive to plant layout than a PID. The distribution of a utility such as steam is by a common distribution pipe, with each unit requiring the utility drawing its supply from that pipe. When a unit has taken its requirement, the utility distribution pipe can then be reduced in size as it continues to the next unit requiring the utility. Clearly the order of servicing units is affected by layout, and utility line sizes are affected by the order.

While serving as a graphical compendium of plant elements, the PIDs really are the most important documents used during project engineering. (Incidentally, a substantial chemical plant may actually require 40–60 related PIDs.) They serve as the basic documents for communication among a variety of specialists whose job is to analyse functional adequacy in relation to control and operability, corrosion and maintenance. Each item is scrutinized to establish that it will perform as desired in the intended service. This document is often used as a basis for contract negotiations between client and engineering firm or division. The PIDs are also the starting basis for vessel design and detailing. Finally, they are used to determine a materials take-off and purchase order list for all standard elements, such as valves and pumps.

At an early stage in their development PIDs and ULDs are issued to enable work on the final stages of plant design to proceed, i.e. the detailed layout and piping design. During this intricate design phase, pipe routes, positions of valves and pipe fittings, and so forth are determined exactly. In recent years it has been common practice to build a model of wire and plastic parts to assist in this exercise and clarify the plant layout and its implications. The application of computers to this aspect of the design process is still in its infancy but a number of systems, notably the PDMS Pipework Design Management System from CADC (1977) and IPAC from IHI (1975) have appeared recently.

From the design work undertaken in this phase, information can be passed on to enable the production of pipework isometrics for use in the fabrication shops and on the construction site, and to enable stock control. These activities are commonly computerized, ISOPEDAC (ICI, 1967) and COMPAID (Davy Computing, 1969) being the most well-known examples.

Another design activity of similar magnitude to piping, and conducted in parallel with it, is the detailed mechanical design of the equipment items such as columns, reactors, pressure vessels and heat exchangers. This is normally undertaken by specialist manufacturers.

The end result of all these interrelated activities is the construction, commissioning and safe operation of a chemical plant. From this admittedly brief discussion one can see that the flowsheeting phase is only a small part of the total effort. Nonetheless, it is the key starting point and how well it is done has significant repercussions throughout the remainder of the design and construction phases, and throughout the operating lifetime of the plant. In other words, the results of the flowsheeting phase are crucial in determining the economics, safety and reliability of the venture.

2

Flowsheeting on the computer

The first fifteen years of history in the development of flowsheeting programs can be broken roughly into three five-year periods. The first developments occurred in the 1955–9 period before computers were of any size and speed, and before languages such as FORTRAN became generally available. The programs which had any degree of success had limited scope, usually to perform the design calculations for a single unit type, for example, heat exchangers. To create larger systems was really too difficult to expect much success.

In the late 1950s Kelloggs (Kesler and Kessler, 1958) began to discuss their Flexible Flowsheet Program. During the next five years (1960–4), several simulators appeared. The successful systems – where 'success' means they survived the period and are still in use – were those which were developed in a high-level language such as FORTRAN, were built in a modular fashion, provided rigorous physical property correlations, and were made robust so that the engineer could almost always get his simulation to work.

The simulators developed were now put to the test of being used on a variety of industrial problems. Many failed, discouraging their users and management which funded them. Thus the next five-year period (1965–9) was a proving and a weeding-out time. Note that large-scale computer power had become generally available by this time.

By 1970 most (but certainly not all) process designers were accepting that flowsheeting programs were useful and cost effective. They were becoming fast enough and reliable enough to be practical for even moderately large processes. Companies buying chemical plants from the contracting companies began to request that the processes be simulated, and, even if the contractor did not prefer to use this tool, he had to use it to satisfy his customer.

The real successes, where considerable engineering time is saved, belong to those, often special-purpose, programs for chemical plants which are repeatedly contracted and built, such as ammonia plants. Sometimes within a day, the calculations can be set up and run, replacing several man-weeks of hand calculations.

We wish in this book to discuss general-purpose flowsheet programs.

In the last chapter we discussed the steps taken in the design effort and some of the relevant computer developments. We shall now attempt first to explain why flowsheeting programs are as they are, discussing in general terms the various approaches which can be taken to write and use these systems.

2.1 Motivation for development

Figure 2.1 gives us a basis for discussing the development of general-purpose flowsheeting programs. The initial programs developed for

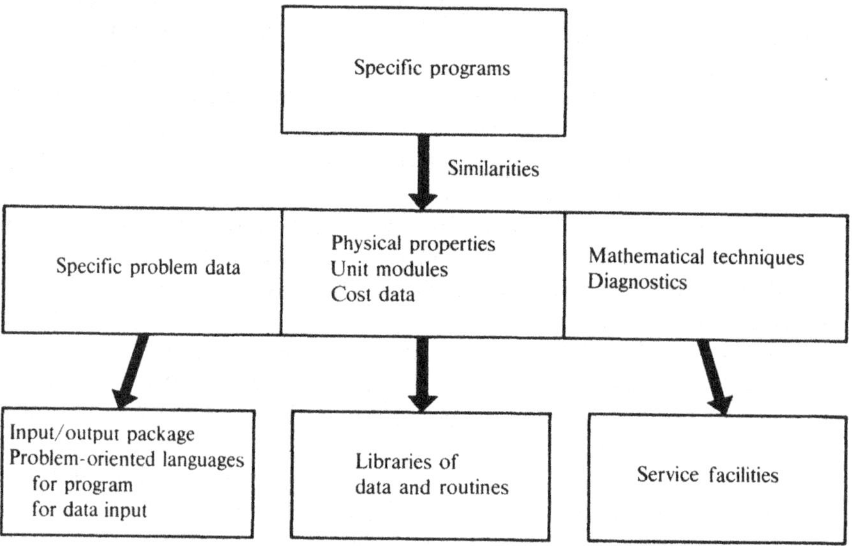

Figure 2.1. Development of simulation software system.

flowsheeting tended to deal with specific examples. The programs were written for a single problem and, even if they were successful, they still made it evident that they were generally too restrictive. The flowsheet could not be readily modified, if at all. A lack of consistency for physical property calculations existed within the program; for example, enthalpies calculated in one unit might not match those found in another. The numerical techniques tended to appear over and over again, for example, to find the root of a single function in one unknown. The diagnostics tended not to be organized.

Advantages existed too. The programs were quite fast. They could be played with until some special-purpose device made them work, even if no theoretical justification could be provided. But, since a principal requirement of design is to try alternatives, the limited, almost nonexistent, ability to make flowsheet changes was far too restrictive.

By the early 1960s companies were building libraries of unit operation programs capable of doing many of the design calculations. It was evident that if these could then be put into a system which could direct the calculations as needed for an entire flowsheet, considerable engineering time might be saved. The experiences of writing these unit operation programs and complete, but special-purpose, flowsheet programs gave the necessary background for the next step.

To design a programming system which would allow performance of the calculations for an entire flowsheet rather than a single unit meant one had to make decisions about a major software system where no one else had: could one make these systems converge for the general case; how much of a numerical analyst would the user have to be; what form of input and output is needed to make the system usable; will people use it?

Examining the special-purpose programs made it clear that the problem could be divided roughly into three parts (see figure 2.1). Part of the problem is very specific to the particular design problem at hand. This includes at the very least the flowsheet, the components list, and the design requirements that the plant must meet. Part of the problem, and hopefully the major part, is common to all problems. The unit operations, types and their calculations should to a large extent be common. The methods to calculate physical properties can hopefully be made common; costing information may also be common. Underlying these, a library of numerical techniques could prove very useful to stop them from being rewritten over and over again, and also to centralize the approach so later improvements might be easy to incorporate. A common approach to diagnostics might also be beneficial.

These observations, viewing figure 2.1, provide the basis for the design of a flowsheeting system which requires a part to handle the problem-specific features, a means to read into the computer a description of these features, and a set of routines that enables the general-purpose parts to be applied in the context of the problem.

A large portion of the system consists of libraries that contain routines for the common calculations. These include a range of unit operations, routines for the generation of physical properties of various chemical compounds and their mixtures, costing routines and sets of mathematical service and diagnostic routines. In addition to the routine libraries there are data libraries (or databanks) that provide basic physical property and costing data for use by the appropriate routines. All the above are problem independent and can be applied as required.

2.2 Developing a simulation model

Figure 2.2 highlights the steps taken to develop a simulation model when using a flowsheeting system. As a first step the user has to prepare and enter the information particular to his flowsheet. He has to list the units and their interconnection. He must supply the list of chemical components in his plant. The design specifications giving certain temperatures, pressures, flowrates, compositions and the like must be entered.

The second step is to take the user information and create a problem definition. The requested physical property methods and support data for the components must be assembled. The calculations desired for the particular units within the plant must be collected. These include heat

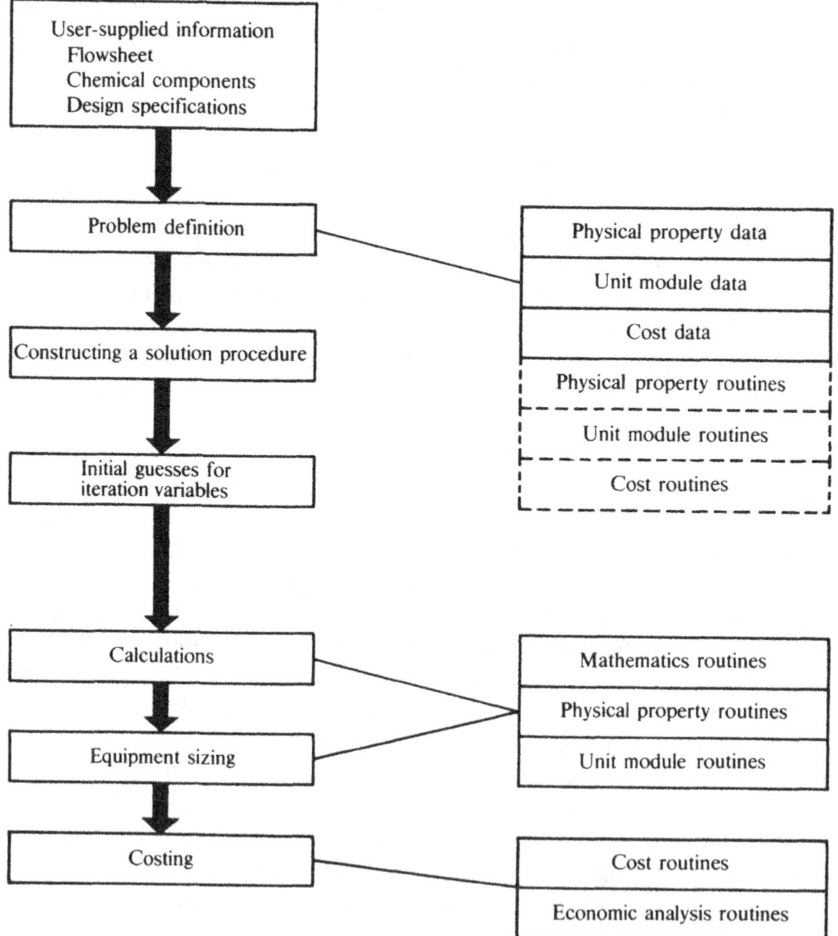

Figure 2.2. Development of a simulation program for a process system.

and material balance equations as well as equipment-sizing type of calculations. Cost calculation methods and their support data must also be brought in. This set of items can get very large very quickly. Also the user, in his enthusiasm, may try to overspecify the problem, which is usually equivalent to giving an inconsistent set of specifications, precluding any solution which fully satisfies them.

Once the problem has been defined, one has to develop a solution procedure to solve it. This, from its most elementary viewpoint, means one has to develop a procedure to solve the hundreds to thousands of equations defining the problem. The complexity of this step varies significantly with the approach one has taken in the problem-definition step. If one is very restrictive in problem definition, then this problem can be fairly simple. In other instances it can become very complex indeed. Later in this chapter we shall see the range of complexity as we discuss some of the alternatives which one can take to create a flowsheeting system.

The actual calculation step of course reflects the approach taken to develop the solution procedure. This step, again because one is solving large sets of nonlinear algebraic equations, faces many problems of a numerical nature. In particular, the system has to get the problem to converge to an answer and, in doing so, it has to decide, in fact, when the problem has converged. It has to overcome, or detect and warn one, when the problem appears to be poorly specified. Chapter 3 will give some insight into these problems.

The final steps in a conventional flowsheeting problem are to perform equipment sizing and costing, and process evaluation.

2.3 Approaches to flowsheeting systems – examples

We shall introduce in this section four different approaches to the development of flowsheeting systems. In later chapters (chapters 6 to 9) three of these approaches plus one additional one will be investigated in more detail. An example problem of a somewhat abstract nature will aid our discussion, and each approach will be discussed with respect to figure 2.2. Many extensions to each of the approaches are possible, and these extensions are often included to overcome the particular deficiencies. Chapters 6 to 9 will discuss some of these extensions in more detail; the discussion here will be limited to the simplest interpretation that one might give to each approach.

The last part of this chapter will compare the approaches, examining qualitatively the relative ease with which one can develop and write the flowsheeting system, and with which the engineer can use the system. Some discussion on program size and computation times will expose further differences.

2.3.1 A simple flowsheet

Figure 2.3 illustrates a very simple flowsheet that we may wish to analyse. It comprises four functional units (two mixers, a reactor and a

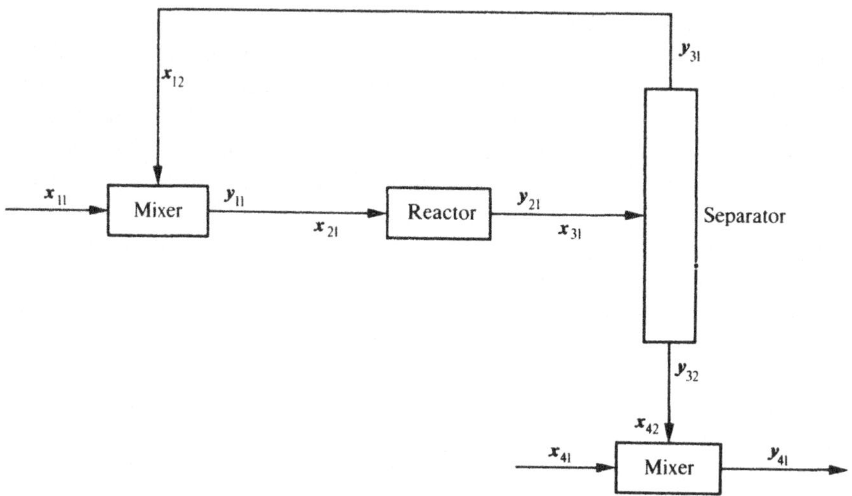

Figure 2.3. A simple flowsheet.

separator) and contains a recycle stream. The equations we can write for this flowsheet are as follows:

(1) *Model equations*

$$\left.\begin{aligned}
f_1(x_{11}, x_{12}, y_{11}) &= 0 \\
f_2(x_{21}, y_{21}, u_2) &= 0 \\
f_3(x_{31}, y_{31}, y_{32}, u_3) &= 0 \\
f_4(x_{41}, x_{42}, y_{41}) &= 0
\end{aligned}\right\} \tag{2.1}$$

using the following notation:

$$\begin{aligned}
x_{ij} \quad &\text{input } j \text{ to unit } i \\
y_{ij} \quad &\text{output } j \text{ from unit } i \\
u_i \quad &\text{unit parameters for unit } i
\end{aligned}$$

(2) *Connection equations*

$$\left.\begin{aligned}
x_{12} - y_{31} &= 0 \\
x_{21} - y_{11} &= 0 \\
x_{31} - y_{21} &= 0 \\
x_{41} - y_{32} &= 0
\end{aligned}\right\} \tag{2.2}$$

The model equations are written functionally and, in fact, represent the heat and material balances, the physical property and other correlations, and so forth, that allow one to relate the unit inputs and outputs. For each unit, a number of equipment parameters u_i also exist which must be specified to complete the unit description. For example, the reactor volume and its operating temperature and pressure are commonly among these parameters. The mixer as considered here has no other parameters, and that is why none are considered in functions f_1 and f_4.

The flowsheet specification for our simple problem can be given to the computer as a list of interconnecting streams. For each unit, then, the input and output streams can be named. The following lists would represent the flowsheet for this problem.

Streams	Link on flowsheet
S1	Feed x_{11}
S2	y_{31}–x_{12}
S3	y_{11}–x_{21}
S4	y_{21}–x_{31}
S5	y_{32}–x_{42}
S6	Feed x_{41}
S7	Output y_{41}

Unit	Model designator	Feed streams	Output streams
1	Mixer	S1, S2	S3
2	Extent-of-conversion reactor	S3	S4
3	Adiabatic flash separator	S4	S2, S5
4	Mixer	S5, S6	S7

This form of data is readily encoded and input to the computer. The model designator is used here to tell the computer which equations the engineer wishes to use to model each unit, as the flowsheeting system could have a number of different separator types or reactor types, each with a different designator and equation set in its library of unit types.

2.3.2 The sequential modular approach

Suppose we make the design decision for our flowsheeting program that all unit models will be written so that one calculates the output stream values of a unit, given its input stream values and the unit parameter values. For a mixer, this requirement is easily seen; by material balance the flow out of each chemical component is readily obtained, and by heat balance we can fix the enthalpy of the outlet stream – since inlet enthalpies must be given. For the reactor, we can use the inputs x_{21}, the functions f_2 as well as values for the unit parameters u_2 to calculate the outputs y_{21}; similarly for the separator. This means functionally that we rewrite the unit model equations as follows:

$$
\begin{aligned}
f_1(x_{11}, x_{12}, y_{11}) &= 0 & y_{11} &= g_{11}(x_{11}, x_{12}) \\
f_2(x_{21}, y_{21}, u_2) &= 0 & y_{21} &= g_{21}(x_{21}, u_2) \\
f_3(x_{31}, y_{31}, y_{32}, u_3) &= 0 \Rightarrow & y_{31} &= g_{31}(x_{31}, u_3) \\
& & y_{32} &= g_{32}(x_{31}, u_3) \\
f_4(x_{41}, x_{42}, y_{41}) &= 0 & y_{41} &= g_{41}(x_{41}, x_{42})
\end{aligned}
$$

Each set of rearranged equations, e.g. $y_{11} = g_{11}(x_{11}, x_{12})$, can be represented by a computer subroutine. For this case its input variables will be x_{11} and x_{12} and its output variables will be y_{11}. The internal calculation would be:

Given as input:	x_{11}, x_{12}
By material balance:	Calculate the exiting flowrate of each chemical component
By heat balance:	Calculate the output enthalpy of the exit stream

These exiting flows and enthalpy are the values represented by the stream values y_{11}.

The second decision we make for designing a sequential modular system is that all input streams (x_{11} and x_{41}) for the flowsheet must be specified by the user. The third decision is that all recycles will be calculated by guessing values for them and iterating on these guesses. For our simple flowsheet in figure 2.3, we might solve the heat and material balance equations as follows:

1. Given input stream values for x_{11}
2. Guess recycle stream values for x_{12}
3. Use model equations $y_{11} = g_{11}(x_{11}, x_{12})$ to calculate y_{11}
4. Equate the stream values for x_{21} and y_{11} (using the appropriate connection equations)
5. Use model equations $y_{21} = g_{21}(x, u_2)$ to calculate y_{21}
6. Equate the stream values for x_{31} and y_{21}
7. Use model equations $y_{31} = g_{31}(x_{31}, u_3)$ and $y_{32} = g_{32}(x_{31}, u_3)$ to calculate the output stream values for y_{31} and y_{32}
8. Examine the difference in stream values for y_{31} and x_{12}, which should be (essentially) equal if we have solved our problem. Then

   (*a*) If not essentially equal, use the values for y_{31}, and x_{12} to guess new values for x_{12}. Repeat from step 3
   (*b*) If essentially equal, the analysis for the recycle has converged. Continue to step 9

9.   Equate stream values for x_{12} and y_{32}
10. Use model equations $y_{41} = g_{41}(x_{42}, x_{42})$ to calculate y_{41}
11. Exit

We have not indicated in step $8(a)$ how to guess new x_{12} values given the y_{31} and old x_{12} values. This topic will be covered in chapters 3 and 6. The latter will give a more detailed discussion of the sequential modular approach.

We can now examine this approach and the use of it to develop a solution procedure as we discussed when considering figure 2.2. We shall briefly discuss each step of the figure.

The user-supplied information is as indicated. The flowsheet needs to be specified and is equivalent to listing each unit and its simulation routine together with how all the units are interconnected. The design specifications are the values for all streams flowing into the process and, for each unit, the equipment parameters u_i. The particular chemical species present will need to be listed.

Given this user-supplied input, the computer is able to reserve storage space for all the streams and for variables calculated by the units listed. It can collect from a thermodynamics databank the needed physical property estimation parameters for the components listed, and finally, for any cost routines listed, it can collect the appropriate cost data.

Constructing a solution procedure may not be required if the flowsheeting system defers this task to the user. He may be required to fix the order in which to calculate the units. The user in this case may have to identify all the recycle streams and explicitly indicate when and how they are to be iterated and converged. For this case, the user must then input the computational sequence for the units, which for our example flowsheet in figure 2.3, might be:

Unit 1, unit 2, unit 3, iterate from unit 1, unit 4

This sequence implies that stream S2, the y_{31}–x_{12} link, is the recycle or 'torn' stream, i.e. the stream which has to be guessed and whose values are iterated. The sequence

Unit 2, unit 3, unit 1, iterate from unit 2, unit 4

is equally possible, and implies stream S3, the y_{11}–x_{21} link, is the 'torn' stream. Note we do not calculate unit 4 until the recycle involving units 1 to 3 are solved completely.

As an alternative, the flowsheeting system can contain algorithms necessary to construct a computational sequence and identify the corresponding torn streams. We discuss this topic further in chapter 6. As the final step for constructing the solution procedure, the various mathematics, unit operations and physical property subroutines are gathered together to make up the computer program which will be executed during the calculation step.

After identifying the torn streams, initial guesses for each must be supplied and these represent the iterates for the flowsheet. In addition, if iterates exist within the computation for a unit, initial guesses for these, too, may have to be supplied. The variables to be guessed are known *when* the unit operations module identifier is specified.

Calculations are simply the execution of the program put together for the problem.

If equipment sizing and costing is done as an integral part of the calculations, the routines to do them are part of the program already constructed. If done separately, a similar sequence of events as that to set up the simulation is needed to set up these calculations.

2.3.3 The simultaneous modular approach

This discussion is to present the essential ideas behind this approach to flowsheeting. Rosen (1962) first discussed a flowsheeting system of this type. Also a system of this type has been developed commercially in Japan (Umeda and Nishio, 1972). The simultaneous modular approach requires all unit modules to be written as they are for the sequential modular approach. Each unit model will be written so one calculates the output stream values of a unit given its input stream values and equipment parameter values. At this point a fundamental difference occurs. A further module for each unit type must also be written which allows one to relate each output value *approximately* to a linear combination of all input values. The calculations to be performed are indicated by the following two-step computation which is to take place for each unit.

1. (*a*) Given the unit inputs $x_{i1}, x_{i2}, \ldots$
   Given the equipment parameters u_i
   Given any initial guesses for iterated variables
   (*b*) Use the unit model to calculate the unit outputs, $y_{i1}, y_{i2}, \ldots$
2. (*a*) Given the unit inputs $x_{i1}, x_{i2}, \ldots$
   Given the unit outputs $y_{i1}, y_{i2}, \ldots$
   (*b*) Find linear relationships which adequately model the units for small perturbations of the inputs, i.e. find the coefficients a_{ijk} such that

$$y_{ij} \simeq \sum_{k=1}^{k_i} a_{ijk}\, x_{ik}$$

where k_i is the number of input streams to unit i.

The complexity of finding the a_{ijk} can vary with the accuracy necessary and desired for these linear relationships if they are to behave as realistic, linear models.

The separator model can serve to illustrate a possible simple linear model for it. Suppose we find after step 1(*b*) above for a column splitting benzene, toluene and xylene that the components in the column split as follows:

Component	Input flow	Top output flow	Bottom output flow
Benzene	100 mol	98 mol	2 mol
Toluene	100 mol	93 mol	7 mol
Xylene	100 mol	9 mol	91 mol
Heat	20 000 kJ	10 000 kJ	10 000 kJ

This represents the output stream values for the exact model using these given input stream values. We might then model the output flows in the simple linear form

$$\text{Benzene top flow out} \simeq 0.98 \ \text{benzene feed flow in}$$
$$\text{Toluene top flow out} \simeq 0.93 \ \text{toluene feed flow in}$$
$$\text{Xylene top flow out} \simeq 0.09 \ \text{xylene feed flow in}$$
$$\text{Heat top flow out} \simeq 0.5 \ \text{heat feed flow in}$$

$$\text{Benzene bottom flow out} \simeq 0.02 \ \text{benzene feed flow in}$$
$$\text{Toluene bottom flow out} \simeq 0.07 \ \text{toluene feed flow in}$$
$$\text{Xylene bottom flow out} \simeq 0.91 \ \text{xylene feed flow in}$$
$$\text{Heat bottom flow out} \simeq 0.5 \ \text{heat feed flow in}$$

For the benzene top flow out the full linear equation for this simple model is

$$\text{Benzene top flow out} \simeq 0.98 \ \text{benzene feed flow}$$
$$+0.00 \ \text{toluene feed flow in}$$
$$+0.00 \ \text{xylene feed flow in}$$
$$+0.00 \ \text{heat feed flow in}$$

and the relevant a_{ijk} coefficients are 0.98, 0, 0 and 0. We note two things. The 0.98 came from observing the results of an accurate model calculation which converted feed stream values into product stream values. The form of the linear model was our decision because with this form we can readily calculate the a_{ijk} values, most of which are zero. A true linearization would not have this simple form. Interactions, represented here by all the a_{ijk} values being nonzero in general, should, in fact, occur. Clearly the input of a hotter feed will affect the top and bottom product compositions. The linear form chosen must represent a compromise between the effort required to find the a_{ijk} values and the accuracy needed to gain, ultimately, the solution to the flowsheet equations.

We should point out that we selected mass and heat flows rather than mole fractions and temperature on purpose as the former are extensive quantities and the latter are intensive. One feels more comfortable stating 90% of a component entering leaves in the top stream than saying 90% of the mole fraction of a component does. Pressure of a stream can also be included. It is an intensive variable requiring some care. Often its value is prescribed throughout a system so it may, in fact, not cause problems.

The remaining equations for the system are the connection equations of the form

$$x_{pq} = y_{st}$$

which are also linear in all variables x and y. After the a_{ijk} are calculated from step 2(b) above we have two sets of linear equations which approximately describe the flowsheet and are of the forms

Approximate models $\qquad y_{ij} \simeq \sum_{k=1}^{k_i} a_{ijk}\, x_{ik}$

Connection equations $\qquad x_{pq} = y_{st}$

We collect these together for the entire flowsheet, that is for all units and all connections. They can, in principle, be solved except that they contain more variables than there are equations. The exact number of extra variables is readily determined because these equations are so similar to the original set, namely

Exact model $\qquad y_{ij} = g_{ij}(x_{i1}, x_{i2}, \ldots, x_{ik_i}, u_i)$

Connection equations $\qquad x_{pq} = y_{st}$

We see that the model equations involve the same x and y variables; only the form has been simplified to give a linear relationship. Also the u_i variables, if fixed in value, are then constants in the exact model. They influence the a_{ijk} coefficients found in the step to fit the linear model to the exact but are not explicitly considered in the linear model. Thus if we choose always to fix the u_i values for the exact model and can devise a calculational sequence in which to solve the exact model, that same sequence should work for the approximate linear model equations too.

In the previous section we devise just such a sequence. It required us to specify the feed stream values. Then the number of equations was sufficient to calculate all values for the rest of the flowsheet. So this observation tells us the number of variables we must specify for our linear equations too.

The principal difference here is that, *because the equations are linear*, we can really make other specifications than the feed streams and still have a fairly easy time solving. (Solving linear equations will be discussed in chapter 3.) These specifications must be equal in number to the number of feed stream variables in our flowsheet, and they must be linear in the x and y variables so the complete approximate model remains linear. Not every combination of specifications is possible, which we shall see later in chapter 5, so one has to be careful to choose a legal set.

We are now in a position to discuss this approach and its use to derive a simulation model as outlined in our earlier discussion in section 2.2 and figure 2.2.

The user-supplied information again includes the flowsheet and a list of chemical components. The design specifications must include all the equipment parameters u_i and a set of linear relationships among the x and y variables equal in number to the number of feed stream variables. They must be judiciously selected so they constitute a set which can lead to a solution.

The problem definition is similar to that for the sequential modular approach. The physical property data, module equation sets, cost data etc. must be gathered together.

Constructing a solution procedure is rather different from the sequential modular approach. The order in which to calculate the units is a more complicated problem because one can place specifications on x and y variables which are not simply feed stream values.

It is possible to analyse the equations within the computer to determine this ordering. In particular, if the feed stream values are specified, the ordering of computations should be the same as for the sequential modular approach. For the moment we shall assume feed stream values are in fact specified and the ordering is as found earlier:

Unit 1, unit 2, unit 3 – as a recycle group

Unit 4   – solved last

The particular sequence of calculations might be the following:

1. Given feed stream x_{11}
2. Guess all a_{ijk} for units 1, 2 and 3. This sounds like a nearly impossible task at first. However if the linear models are kept reasonably simple, the a_{ijk} values can often be guessed without too much difficulty
3. Solve the linear form for the model equations together with the connection equations for units 1, 2 and 3. This step will give approximate values for all the x and y variables for these three units
4. Using the x-variable values just found as the unit feeds and the given equipment parameters u_i, solve the exact model equations for unit output variable values for units 1, 2 and 3
5. Recalculate the a_{ijk} coefficients for each unit

6. Compare the new a_{ijk} coefficients to the ones previously guessed. Then
   (*a*) If essentially equal, continue with step 7
   (*b*) If not essentially equal, reguess their values and repeat from step 3
7. Since x_{41} is a given feed to unit 4 and x_{42} was just calculated as the output of unit 3, unit 4 can be calculated directly. Use the exact model equations to solve.
8. Exit.

Steps 2 and 3 require further discussion as they may seem to be rather difficult. For one thing the number of a_{ijk} coefficients is large. But if the linear models adopted are simple, most will be zero. If too simple, this algorithm may not converge.

It is also possible to guess stream values, for example the inputs to all units, or to reduce this number by guessing only recycle stream values. We can consider the resulting algorithm for this last approach for our sample problem.

1. Given feed stream x_{11}
2. Guess the recycle stream values only for x_{12}
3. Using the sequential modular approach solve units 1, 2 and 3 in that order, using as inputs to unit 2 the output values of unit 1 and as inputs to unit 3 the just calculated outputs of unit 2
4. Find the a_{ijk} coefficients for units 1, 2 and 3
5. Solve the linear form for the model equations together with the connection equations for units 1, 2 and 3. This step will give new values for all x and y values except those of the given feed x_{11}
6. Compare the recycle stream values x_{12} just calculated to those guessed. Then
   (*a*) If not essentially equal, reguess and repeat from step 3
   (*b*) If essentially equal, go to step 7
7. Solve unit 4 using the exact model equations
8. Exit

At this point it is obvious that the choice of which values to guess and to iterate (i.e. to tear) is not unique. If the feed streams are not specified, but a rather more general (but linear) specification is provided in terms of other x and y values, the sequence of calculations is different and must be worked out by the computer or by the user. A default option is simply to solve *all* units each iteration. This option may be costly. For example in the above case, unit 4 calculations had to be done once only after those for units 1, 2 and 3 were complete. If unit 4 was a complex unit (or sequence of units), not just a simple mixer, this default is to calculate it inside the iteration loops for units 1, 2 and 3 and this could be very costly.

In any case, once the algorithm to solve is fixed, the variables, which are the iterates, are all identified and values will have to be entered. Again iterates within the unit model routines are also to be included in the set of iterates for the flowsheet.

The calculation portion is to execute the above procedure, which is in the form of a computer program put together in the previous steps.

Sizing and costing calculations are as discussed in the sequential modular case.

2.3.4 Equation-solving approaches

In the previous two sections we found that each unit i was modelled by writing a computer subroutine which could convert input stream, x_{ik}, and equipment parameter, u_i, values into output stream values, y_{ij}. We now examine the implications of removing this requirement.

The final flowsheet is represented by a collection of nonlinear equations which must be solved simultaneously. The equations included are the model equations and connection equations, for our sample problem, equations (2.1) and (2.2). The design specifications are in fact another set of equations. For example, to require that a certain flowrate F in our problem be set to 300 kgmol/h is equivalent to requiring the equation

$$F - 300 = 0$$

to be satisfied. The temperature T_{15} in one part of a flowsheet might have to be 30 °C greater than in another, T_6

$$T_{15} - T_6 - 30 = 0$$

In the last two methods described, it was generally required that all equipment parameters needed for the model equations had to be specified. Often, if not usually, one wishes to specify many of the values for both unit inputs and outputs. The unit equipment parameters are then calculated to give these desired transformations of inputs to outputs by the unit; in other words, the unit is *designed* to meet these requirements. This distinction is that which differentiates 'rating' calculations from 'design' calculations. For rating, the unit is specified and its performance is calculated. For design, the performance is specified and the unit is designed to meet this performance, if it can. In the equation-solving approach this is feasible. The requirements are written as equations and added to the set defining the problem. If they represent a legal specification set, then they should create no problem, in principle.

Once the full set of equations are together, a solution procedure for them is derived automatically by the computer. This effort is considerably more complex because of its large size than having the computer find a solution procedure for the simultaneous modular approach. However, many of the underlying ideas are similar.

If the solution procedure is examined, one will find unspecified equipment parameters are often the output of a calculation. Thus, instead of heat duty on an exchanger being input, it may be calculated. In the next exchanger it may indeed be input because the user wishes to specify it.

We can now discuss, as we have for the previous methods, the use of this approach for developing and using a simulation model. Again we shall look at the steps listed in figure 2.2.

The user-supplied information is the flowsheet and the list of chemical components for the process. The design specifications are in the form of a set of (nonlinear if desired) equations which may be fairly general in their structure.

The problem definition step is the collecting together of the equations implied by the units in the flowsheet. These will be kept in a library as equation sets. For the two modular approaches they were kept as subroutines. The flowsheet connection equations will be added as are the user-supplied design specifications. This collection could comprise thousands of equations.

The constructing of a solution procedure means the flowsheeting program has to do the equivalent of writing a large computer program to solve thousands of equations. The approaches possible will be discussed in chapter 3, and they include procedures based on both 'tearing' (see Leigh, Jackson and Sargent, 1974; and Westerberg and Shah, 1978) and simultaneous linearization (see Hutchison and Shewchuk, 1974; and Kubicek, Hlavacek and Prochaska, 1976).

The calculations section is to execute, as before, the solution procedure just set up. Sizing and costing can be handled as a further part of the simulation and can use equation-solving or modular ideas. If a decision is to *supply* the cost of a unit as an input – not a totally unreasonable idea – then the sizing and costing equations could be included above so that the solution procedure can account for receiving cost as an input rather than producing it as a result of the calculations.

2.3.5 A comparison of the approaches

The approaches given in the last section are only a sampling of the variety of approaches one can visualize for a flowsheeting system. They range from methods which require the user to be very rigid in his problem definition to ones which allow him to be very flexible. We shall now discuss some of the implications of these differences on the creation of a flowsheeting system and its use by the design engineer. As with all things, nature has extracted its toll. The easier it is to build a system, the more difficult it is for the design engineer to use it for a number of common problems he may wish to solve.

The first topic we shall examine is the ease with which one can write the flowsheeting executive system for each of the approaches. The flowsheeting executive is the program which receives the user input, collects the problem description together, etc., right through to executing and costing a system.

The sequential modular executive is the easiest to write. The assumption that each unit operation module will calculate unit output stream values given unit input stream values and equipment parameter values solves many problems for the system executive system. In contrast, the equation-solving executive can become extremely complex. The problems are those of guaranteeing that the design engineer will give a legitimate problem definition to the system and that the solution procedure used by the system can be made to converge to an answer.

By assuming a rigid form for writing a unit module as in both modular approaches, a subroutine for each module can readily be written ('readily' is a comparative term here). The writer has to make no decisions on the type of input data that he will be given. The preparation of a unit module for an equation-solving approach, in fact, is not much more difficult, and may be easier. The writer has to gather together the equations defining the unit into a form that the system can recognize and handle.

The executive for the sequential modular approach has to analyse the flowsheet, if it is designed so that the user does not have to do this step, and work out a computational order for the unit modules. The literature abounds with methods which will solve this problem and, since a flowsheet comprises relatively few units and streams, most of the methods can give good answers quickly. The combinatorial problem of looking at most orderings does not have to be an excessive one even for a fairly large system.

The executive for the simultaneous modular system also has to analyse the flowsheet, but it has to decide only about which units are involved in a computational loop. This loop does not follow the flow of material through the flowsheet since the user can give specifications on output stream values as well as input stream values. The units in a computational loop are solved in a parallel fashion so that the ordering within a group is not required. The connecting of the units is one of solving the approximating linear equations. These equations contain a computational loop, in principle, but, being linear, they are readily solved by so-called direct methods.

For an equation-solving approach two options occur. The mass of thousands of equations can be solved by successive linearization methods, or they can be approached by an analysis which attempts to minimize the number of iterate variables needed, a method called 'tearing'. In the former version, the system must calculate or estimate needed partial derivatives, literally thousands of them, and set up a set of linear equations to solve simultaneously at each iteration. The tearing approach can be very complex. The resulting solution procedure has to converge, for example. It represents a combinatorial problem two-or-three orders of magnitude more difficult to solve than reordering a flowsheet for the sequential modular approach. Here the exact use to be made of each and every equation and where it is to occur in the computational sequence must be determined. The choice of which other decision variables the user might usefully specify to aid the computation has to be fixed if the user has not given a sufficient number of specifications.

We can see then that the section in figure 2.2 titled 'construct a solution procedure' can vary from quite straightforward to very complex. Since the complex form of this problem is not yet fully solved in the literature, it is not surprising that almost all existing flowsheeting systems have avoided it and have adopted the sequential or simultaneous modular approaches.

The cost for avoiding it is that a general specification, which will then require an equipment parameter to be calculated, must be handled indirectly and the computation time in the step called 'calculations' is likely to be much larger. The advantage of avoiding it is that the flowsheeting system can be designed and written much more quickly, the system is simple to specify and the unit modules can be written, tested and made computationally robust by taking special-purpose precautions.

The design engineer also has no problem knowing what he has to specify for the sequential modular approach to make a well-defined problem. Thus data checking is much less difficult for him and for the system too.

To overcome the requirement that equipment parameters must always be specified, special unit modules can be built into the sequential modular approach, and we shall discuss these in more detail in chapter 6. Also the computation times can be reduced by other clever devices, such as not converging to a solution within the module calculations until the recycle computations are also close to convergence. We shall discuss these ideas too in chapter 6.

3

Solving linear and nonlinear algebraic equations

Underlying a flowsheeting system must be some service routines which can be used to solve equations numerically. The simplest problem of practical need is to solve a single nonlinear equation in one unknown. We shall examine this problem first and develop a common simulation problem where this arises. Although the number of strategies to solve this problem are numerous, we shall concentrate on a single method and develop it sufficiently so that it can handle some poorly behaved problems.

The next general service routine is one to solve sets of linear equations. Since most techniques to solve nonlinear equations require an iterative solution of linear equations, this routine is very generally useful. The stress again is the practical requirements for a flowsheeting system. Here the sets of linear equations are often very large (several thousand may arise) and very sparse. By 'sparse' we mean that each equation will have few nonzero coefficients. For example an equation

$$\sum_{j=1}^{1000} a_{100,j}\, x_j = b_{100}$$

appears to have 1000 coefficients $a_{100,j}$. For the problem at hand it may be a simple equation

$$10x_{50} - 15x_{75} + x_{512} = -3$$

which in fact has three nonzero coefficients and 997 zero-valued coefficients. We shall cover the practical implications of this sparseness on solving linear equations.

The third section will discuss the methods one might use to solve sets of nonlinear equations. These too are sparse for almost all real problems and two methods have been developed in the literature: tearing and successive linearization. We shall investigate these approaches with practical engineering examples.

3.1 Solving one equation in one unknown

This problem sounds simple, yet the literature abounds with papers

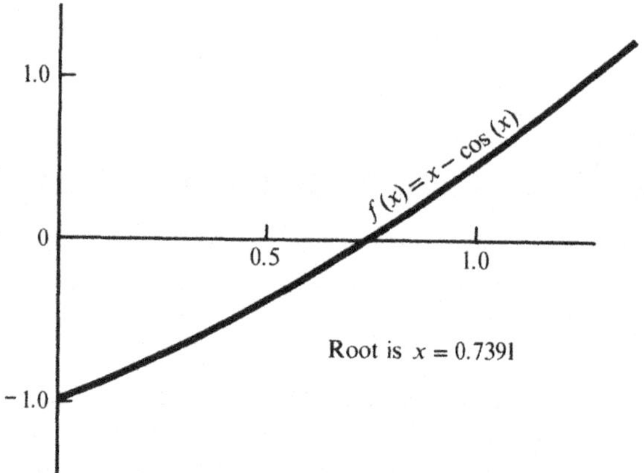

Figure 3.1. Finding the root of a single function.

aimed specifically at solving it. The problem of finding x such that

$$f(x) = x - \cos(x) = 0; \; x \text{ in radians} \tag{3.1}$$

is shown graphically in figure 3.1. A reasonable way to solve is to plot $x - \cos(x)$ versus x and find where it is zero, at about $x = 0.74$ if one plots it carefully. Of course this approach is not useful for the computer. If we assume the derivative of $f(x)$ is available, we can expand $f(x)$ linearly in a Taylor series about the current point $\hat{x}$,

$$f^*(x) \simeq f(\hat{x}) + \left(\frac{\mathrm{d}f}{\mathrm{d}x}\right)_{\hat{x}} (x - \hat{x}) \tag{3.2}$$

$f^*(x)$ is the linear function which has the same value and first derivative at $x = \hat{x}$ as $f(x)$, and is illustrated in figure 3.2. We can find the solution

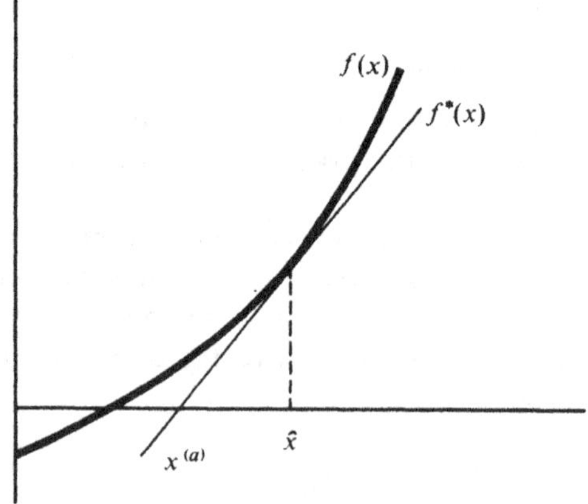

Figure 3.2. A linear function $f^*(x)$ which approximates $f(x)$ around $x = \hat{x}$.

to $f^*(x) = 0$ as it is linear in x:

$$x^{(a)} = \hat{x} - \left(\frac{df}{dx}\right)_{\hat{x}}^{-1} f(\hat{x}) \tag{3.3}$$

$x^{(a)}$ is shown in figure 3.2. We can use $x^{(a)}$ as our estimate for the root of $f(x)$. This approach is known as the Newton method for finding the root of a function. Applying this idea to our earlier problem, we get the following results with an initial guess for $\hat{x}$ of 0.5:

$$f(x) = x - \cos(x) \tag{3.4}$$

$$\frac{df}{dx} \equiv f'(x) = 1 + \sin(x) \tag{3.5}$$

$\hat{x}$	$f(\hat{x})$	$f'(x)$	$x^{(a)}$
0.5000	−0.3776	1.4794	0.7552
0.7552	0.0271	1.6855	0.7391
0.7391	0.0000	1.6737	0.7391

In two iterations four digits of accuracy are found for the answer. It is obvious from looking at figure 3.1 that $f(x)$ is well approximated by $f^*(x)$, and this quick convergence is a result.

One commonly cannot or is not willing to do the algebra to evaluate df/dx functionally, as we did in (3.5) for our simple example. For this case, we can approximate it by evaluating $f(x)$ at two points, x_1 and x_2:

$$\frac{df}{dx} \simeq \frac{f(x_2) - f(x_1)}{x_2 - x_1}$$

Then equation (3.3) becomes the commonly used secant approximation to our root (letting x_2 be $\hat{x}$).

$$x^{(b)} = x_2 - \left(\frac{f(x_2) - f(x_1)}{x_2 - x_1}\right)^{-1} f(x_2) \tag{3.6}$$

Applying this to our problem we get the following results with initial values for x_1 and x_2 of 0.5 and 0.6, respectively.

x_1	x_2	$f(x_1)$	$f(x_2)$	$x^{(b)}$
0.5	0.6	−0.3776	−0.2253	0.7480
0.6	0.7480	−0.2253	0.0150	0.7388
0.7980	0.7388	0.0150	−0.0005	0.7391
0.7388	0.7391	−0.0005	0.0000	0.7391

In applying the secant method we note that five distinct function evaluations (underlined for emphasis) were needed. After the first line, we simply make x_1 our previous x_2 so that $f(x_1)$ need not be re-evaluated. We note that, if evaluating $f'(x)$ in the Newton method is comparable in effort to evaluating $f(x)$, then the secant method has been about as efficient, and it usually is.

We observe some interesting questions if we think about these calculations. First, how did we decide to stop iterating? Clearly two clues are available. The first is that the root guessed, $x = 0.7391$, was exactly the next root calculated to four decimal places. Also we see that $f(0.7391)$, to four decimal places, is zero. Neither clue tells us the whole story. Figure 3.3 gives two instances when either could be true, but not both.

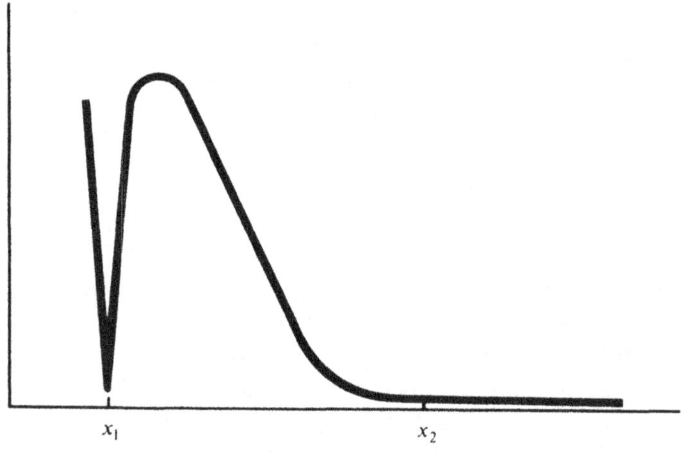

Figure 3.3. Difficulties with numerical root finding.

A Newton iteration in the vicinity of x_1 will find $x^{(a)} \simeq x_1$ and one near x_2 will find $f(x) \simeq 0$. Neither is a root or even near to a root. Also, at $x = x_1$ exactly, $f'(x)$ is zero (if $f'(x)$ is continuous) and will cause us to divide by zero in equation (3.3).

If we can guarantee $f(x)$ is continuous then we can apply another type of technique to find x, a type of area elimination. If we can find two points $x^{(1)}$ and $x^{(2)}$ where $f(x^{(1)})$ is of opposite sign to $f(x^{(2)})$, then we know a root lies between $x^{(1)}$ and $x^{(2)}$. Figure 3.4 illustrates this.

We can eliminate half the search area by a single function evaluation at the midpoint of the interval, x_m. Depending on the sign of $f(x_m)$, we eliminate the area to its right or to its left and repeat. To get an accuracy of 0.1% of the original interval size requires 12 function evaluations. Thus it converges much less rapidly than our previous methods but it must converge regardless of how miserable $f(x)$ is between $x^{(1)}$ and $x^{(2)}$. It does require one to find an interval surrounding the root; this is an additional penalty.

A reasonable algorithm which will do quite well on many problems can incorporate both the secant and area-elimination ideas. Let x_l be the lower bound on x and x_u the upper bound. The secant method can be used unless it fails, meaning it predicts repeatedly that the next root is outside the allowed bounds. If the root is bounded (that is, $f(x_u)$ has the opposite sign to $f(x_l)$) then x can be set to the midpoint value.

If not bracketed, a search can be started. Whenever the secant method starts to give encouraging results again, it can be re-used. Once

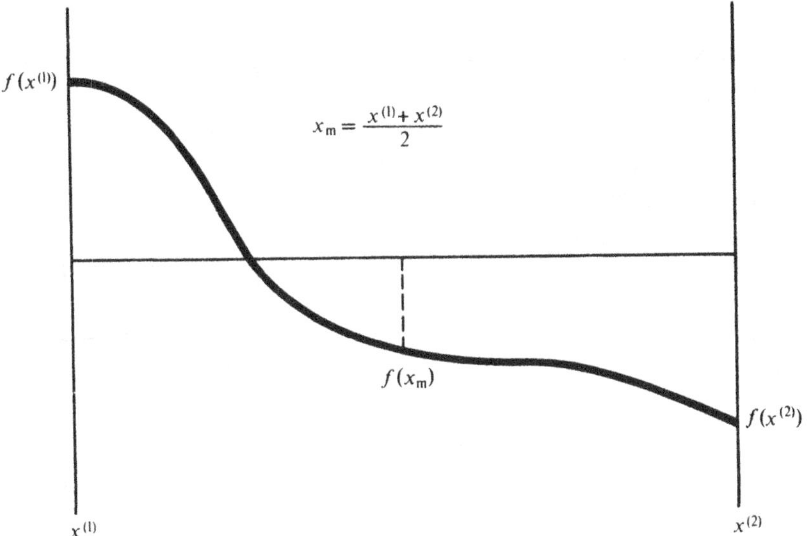

Figure 3.4. Half-interval-area elimination (binary search) to locate a root.

the root is bracketed, every new function evaluation allows the bracketing to be improved. Figure 3.5 illustrates this approach on an unpleasant function.

The original bounds as well as guesses in the form of $x^{(1)}$ and $x^{(2)}$ are provided by the user. The secant method predicts $x^{(3)}$ below the lower bound so it is placed at the lower bound instead. At this point $f(x^{(3)})$ is found to be of opposite sign to $f(x^{(1)})$ so the upper bound is moved to $x^{(1)}$. Points $x^{(2)}$ and $x^{(3)}$ predict $x^{(4)}$ by the secant method; again the upper bound can be lowered. Points $x^{(3)}$ and $x^{(4)}$ predict $x^{(5)}$, again lowering the upper bound. Points $x^{(4)}$ and $x^{(5)}$ predict below the current lower bound $x^{(3)}$ and $f(x^{(3)})$ is already known. Therefore, $x^{(6)}$ is placed at the midpoint of the current bounds, i.e. halfway between $x^{(3)}$ and $x^{(5)}$. At $x^{(6)}$ the lower bound can again be moved. Points $x^{(5)}$ and $x^{(6)}$ predict $x^{(7)}$. At this point the secant method will predict the root very accurately and convergence will be fast.

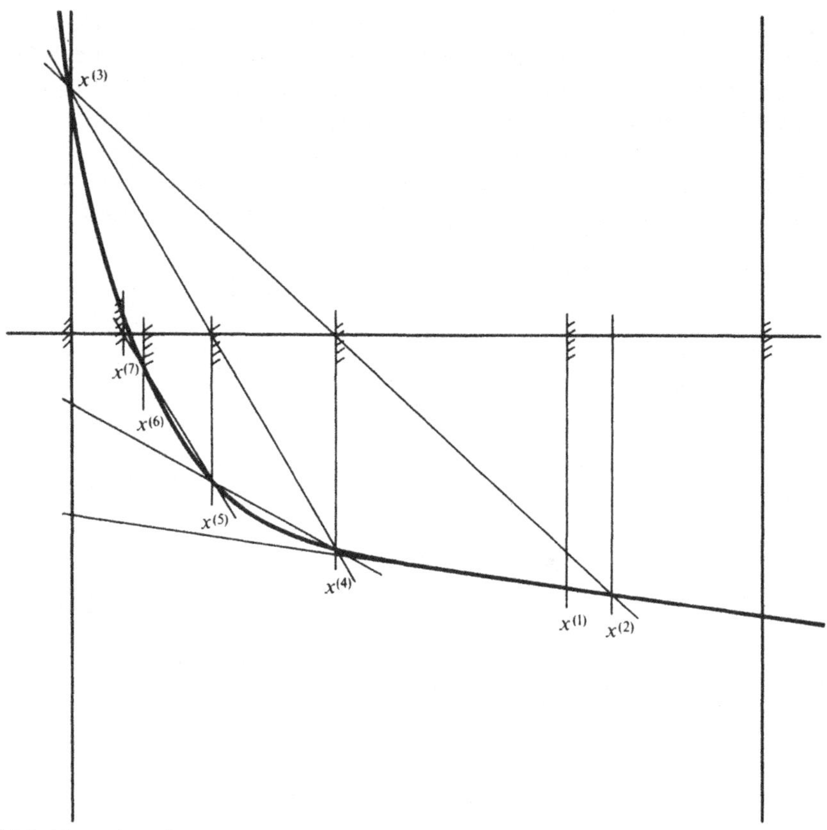

Figure 3.5. Application of a secant/half-interval method.

3.1.1 A flash calculation

We can demonstrate the usefulness of a root-finding procedure on a simple problem repeatedly faced by an engineer when doing design calculations. The problem will be a simplified isothermal flash calculation; see figure 3.6. The equations we shall use to model this unit are as follows.

Material Balance:

$$y_i V + x_i L = z_i F \quad i = 1, 2, \ldots, c \tag{3.7}$$

$$V + L = F \tag{3.8}$$

Equilibrium:

$$y_i = K_i x_i \tag{3.9}$$

Physical Properties:

$$K_i = P_i^0/P \qquad (3.10)$$

$$P_i^0 = 10^{A_i - B_i/(C_i + T - 273.15)} \quad A_i, B_i, C_i \text{ given} \qquad (3.11)$$

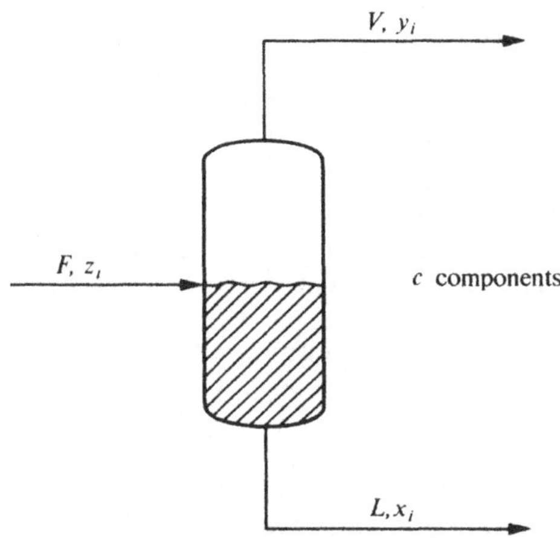

Figure 3.6. A simple flash calculation.

Other:

$$\Sigma x_i - \Sigma y_i = 0 \qquad (3.12)$$

$$\Sigma z_i = 1 \qquad (3.13)$$

Specifications:

$$z_i(i = 1, 2, \ldots, c-1), P, T, F \qquad (3.14)$$

where the notation is as follows:

x_i, y_i, z_i	mole fractions for component i
V, L, F	molar flowrates
K_i	vapour–liquid equilibrium K-value for component i
P_i^0	vapour pressure of component i
P	total system pressure
T	absolute temperature
A_i, B_i, C_i	constants for Antoine vapour pressure equation for component i

Equation (3.10) makes it evident that we are assuming that equilibrium is defined by Raoult's law.

By combining equations (3.7) and (3.9) to eliminate all y_i variables, we get

$$(K_iV+L)x_i = z_iF \quad i = 1, 2, \ldots, c \tag{3.15}$$

which then allows us to see our way to a solution procedure.

1. Given that values are available for $z_i (i = 1, 2, \ldots, c-1)$, P, T, F
2. (a) Using equation (3.13): $z_c = 1-(z_1+z_2+ \ldots +z_{c-1})$
   (b) Using equation (3.11): $P_i^0 = 10^{A_i-B_i/(C_i+T-273.15)}$
   (c) Using equation (3.10): $K_i = P_i^0/P, \quad i = 1, 2, \ldots, c$
3. Guess vapour fraction V/F; V then equals $(V/F) \cdot F$
4. (a) Using equation (3.8): $L = F-V$
   (b) Using equation (3.15): $x_i = z_iF/(K_iV+L) \quad i = 1, 2, \ldots, c$
   (c) Using equation (3.9): $y_i = K_ix_i \quad i = 1, 2, \ldots, c$
5. Evaluate equation (3.12): $\sum_i x_i - \sum_i y_i$

   (a) If not essentially zero, reguess V/F and repeat from step 4
   (b) Otherwise, continue with step 6
6. Exit

Investigating steps 3 to 5, we see that we have, in fact, been finding the root of a single nonlinear function in a single unknown. The unknown is the vapour fraction V/F and the function we wish to drive to zero is $\sum x_i - \sum y_i$. An isothermal flash of a mixture of 100 mol of n-pentane, n-hexane and n-heptane produces the plot of $\sum x_i - \sum y_i$ shown in figure 3.7, where the first five steps of the bounded secant method are

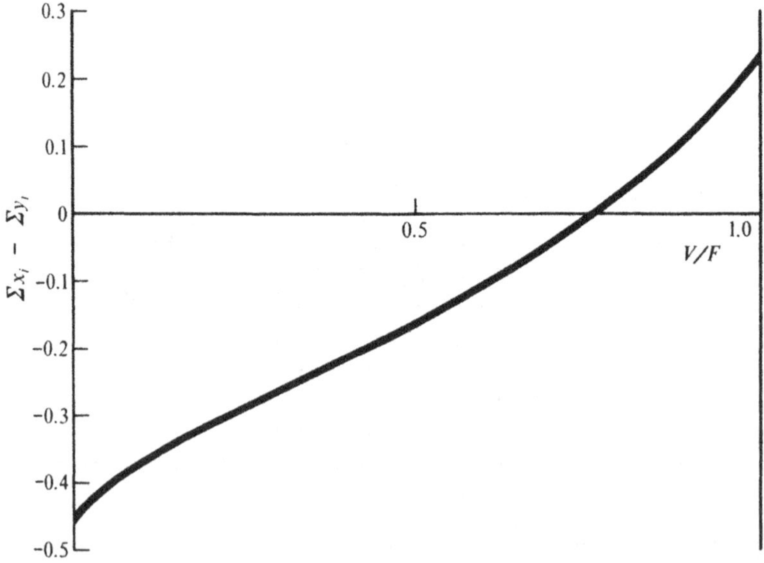

Figure 3.7. Plot of error function for simple flash calculation specified in table 3.1.

illustrated.   Initial   guesses   of $(V/F)^{(1)} = 0.5$ and $(V/F)^{(2)} = (V/F)^{(1)} + 0.05$ were used. Table 3.1 gives the problem specification and table 3.2 the results of the secant search.

Table 3.1 *Given data for simple isothermal flash problem*

$F = 100$ mol		$T = 330$ K		
$z_1 = 0.65$		$P = 760$ mmHg		
$z_2 = 0.2$				
Component	Name	A_i	B_i	C_i
1	n-Pentane	6.85221	1064.63	232.000
2	n-Hexane	6.87776	1171.530	224.366
3	n-Heptane	6.90242	1268.115	216.900

A_i, B_i and C_i values were obtained from Dean (1973).

Table 3.2 *Points evaluated during secant search for root for flash unit example*

i	$V/F^{(i)}$	Error
1	0.50000	-0.15397
2	0.55000	-0.12802
3	0.79666	0.02358
4	0.75829	-0.00436
5	0.76428	-0.00016
6	0.76450	0.00000

This problem can become considerably more complicated. For example, the temperature and pressure specifications may result in no root between $V = 0$ and $V = F$. In this case the mixture is single phase. Also the physical properties, i.e. the K-values, may be the result of nonideal correlations so they are also functions of compositions. In this case, no solution procedure can be worked out which is based on guessing a single variable. The usual procedure for detecting whether the mixture is single phase is to check its dew and bubble points to see if the temperature specified is between them. We shall discuss the nonideal situation later when we are better prepared to handle several nonlinear equations in several unknowns.

3.2 Solution methods for linear equations

This section is not intended to give a complete review of methods of solution for linear equations. It will describe only methods which have been extensively applied in the context of steady-state simulation, with the elementary material necessary for the understanding of the methods. Further discussion is to be found in Fox (1964), for example.

3.2.1 Gauss elimination

Methods for the numerical solution of linear equations fall into two classes, direct methods and iterative methods. Iterative methods, such as the Jacobi and the Gauss–Siedel methods, are generally useful only when one coefficient is substantially larger than the other coefficients in each equation, and the equations can be reordered so that the dominant coefficients lie on the diagonal of the coefficient matrix. These conditions are not usually met in the context of steady-state simulation, and these iterative methods will not be further considered in the solving of linear equation sets. We shall, however, consider related iterative methods in detail in section 3.3.2 for solving nonlinear equation sets such as those arising in flowsheeting.

For sets of linear equations in which the matrix of coefficients is not symmetrical, the best direct method is Gauss elimination. This is the familiar method of elementary algebra put into systematic form and with certain refinements.

In order to fix attention on a particular case, consider a set of four equations in four unknowns, a problem not too big to be tackled manually with the aid of a calculator, but large enough to justify setting it out systematically. With a desk or pocket calculator it is advisable to make certain preliminary adjustments to the equations to get them in the best form for solution. These adjustments take the form of scaling. Any equation can be scaled without difficulty by multiplying throughout by any desired factor. Any variable can be scaled down by multiplying all the coefficients of that variable by any desired factor; it is necessary then, when the solution process is complete, to scale up by multiplying the variable by the same factor. Scaling should be done, if necessary, so that the largest term in each equation has a magnitude of about 0.5, and so that no variable has all its coefficients substantially less than this value. In computer work scaling is usually unnecessary.

Suppose then that the equations (3.16) have been appropriately scaled.

$$
\left.
\begin{array}{l}
0.2246x_1 + 0.3860x_2 + 0.4050x_3 + 0.1220x_4 = 0.2720 \\
0.4082x_1 + 0.1432x_2 + 0.3713x_3 + 0.3005x_4 = 0.4713 \\
0.3672x_1 + 0.1896x_2 + 0.3913x_3 + 0.0640x_4 = 0.5659 \\
0.1702x_1 + 0.4021x_2 + 0.2732x_3 + 0.3801x_4 = -0.1794
\end{array}
\right\} \quad (3.16)
$$

The first part of the solution process can be described in words as follows:

1. (*a*) From the first remaining column of coefficients, select the largest element in magnitude. Call it the pivot element, and its equation the pivot equation

    (*b*) For each of the other equations, establish that multiple of the pivot equation which when added to the equation will reduce its first coefficient to zero. Record the necessary multiplier

(c) Using the multipliers established in 1(b), form new equations by addition of the appropriate multiple of the pivot equation to each of the remaining equations. This operation reduces the number of equations by one, eliminating at the same time one variable

(d) Repeat from 1(a) until one equation remains in one unknown

This part of the process is sometimes known as triangularization, because if the pivot equations are set out in order in an array at this stage it is a triangular array.

Applying this process to the equations (3.16) (more conveniently set out in table 3.3), the element 0.4082, the largest element in the column of coefficients of variable x_1, is selected as pivot, and equation 2 is selected as pivot equation. The multipliers -0.55022, -0.89955 and -0.41695 are selected so that these multiples of equation 2, added to equations 1, 3 and 4 respectively, reduce the coefficient of x_1 in the new equations to zero. The new equations are formed and recorded as a block of three equations in three unknowns with the variable x_1 eliminated. Repetition of the process from 1(a) leads to the selection of the element 0.34239 in the column of coefficient of variable x_2 as pivot and the new equation 3 as pivot equation.

Table 3.3. *Gauss elimination – triangularization*

Multiplier	x_1	x_2	x_3	x_4	RHS	Check sum
-0.55022	0.2246	0.3860	0.4050	0.1220	0.2720	1.4096
	<u>0.4082</u>	0.1432	0.3713	0.3005	0.4713	1.6945
-0.89955	0.3672	0.1896	0.3913	0.0640	0.5659	1.5780
-0.41695	0.1702	0.4021	0.2732	0.3801	-0.1794	1.0462
-0.89725		0.30721	0.20070	-0.04334	0.01268	0.4772
-0.17751		0.06078	0.05730	-0.20631	0.14194	0.0537
		<u>0.34239</u>	0.11839	0.25481	-0.37591	0.3396
			<u>0.09447</u>	-0.27197	0.34997	0.1724
-0.38403			0.03628	-0.25154	0.20867	-0.0065
				<u>-0.14710</u>	0.07427	-0.0728

Selected pivots are underlined.

Notice that the multipliers are set out in a special column to the left of the main block, and that a check sum is maintained on the right. This check sum is, of course, not necessary in computer work, but is important in manual computation. The check sum for an original equation is the sum of all the coefficients and the right-hand side (RHS) element of that equation. The check sum for subsequent equations is obtained by performing operations on the check sums as if they were constituent terms of the equations, and should agree (to within one-or-two decimal digits) to the sum of the terms in the equations.

The second part of the process of solution can be described in words as follows, working on the triangular array of pivot equations (3.17):

$$\left.\begin{array}{rl}0.4082x_1 + 0.1432x_2 + 0.3713x_3 + 0.3005x_4 &= 0.4713\\ 0.34239x_2 + 0.11839x_3 + 0.25481x_4 &= -0.37591\\ 0.09447x_3 - 0.27197x_4 &= 0.34997\\ - 0.14710x_4 &= 0.07427\end{array}\right\}(3.17)$$

2. (*a*) Solve for the variable corresponding to the last remaining column by dividing the RHS element by the coefficient of the variable. Call this result the current solution
    (*b*) Use the current solution to eliminate this variable from the remaining equations
    (*c*) Repeat from 2(*a*) until there are no remaining equations

This part of the process is known as backsubstitution. Applying this process to the equations (3.17) (more conveniently set out in table 3.4) the result for x_4 (−0.50489) is obtained immediately by dividing the RHS element by the coefficient of x_4 in equation 4. This result is immediately used to create new RHS elements by subtracting the product of this current result and the coefficient of x_4 in equations 1, 2 and 3 from the existing RHS elements. The process is then repeated to obtain the result for x_3. The full result is shown in table 3.4.

Table 3.4. *Gauss elimination – backsubstitution*

x_1	x_2	x_3	x_4	RHS	Results
0.4082	0.1432	0.3713	0.3005	0.4713	
	0.34239	0.11839	0.25481	−0.37591	
		0.09447	−0.27197	0.34997	
			−0.14710	0.07427	−0.50489
0.4082	0.1432	0.3713		0.62302	
	0.34239	0.11893		−0.24726	
		0.09447		0.21266	2.25103
0.4082	0.1432			−0.21278	
	0.34239			−0.51376	−1.50050
0.4082				0.00209	0.00512

Quantities with overbar need not be written into the table.

In manual computation, as a check that blunders have not occurred, it is usual to verify that the results obtained satisfy the sum of the equations (3.16), i.e. equation (3.18), to the accuracy implied by the coefficients:

$$1.1702x_1 + 1.1209x_2 + 1.4408x_3 + 0.8666x_4 = 1.1298 \quad (3.18)$$

An important modification of this method is known as the Gauss–Jordan method. In this method, when a pivot has been chosen, it is used to eliminate the elements in its column from the previous pivot equations as well as from the as yet unused equations. There is however, substantial disadvantage in the method as it usually requires about twice as many arithmetic operations as a Gauss elimination with backward substitution to solve a set of linear equations.

Application of the process, partly completed, is set out in table 3.5, which illustrates its use on the same original equations. The process as regards the first pivot is exactly as described previously. When the pivot 0.34239 is selected for the second column, equation 4 is used to eliminate the second-column terms from all three equations. Note that when 0.09447 is selected in the third column, the choice is restricted between that and 0.03627, as equations 1 and 3 are the two remaining unused equations. Note that this process terminates when there are four isolated coefficients of four distinct variables on the left-hand side (LHS) of the equations, with corresponding RHS elements, so that the solutions are finally obtained by simple division.

Table 3.5. *Gauss–Jordan elimination, partially completed*

Multiplier	x_1	x_2	x_3	x_4	RHS
−0.55022	0.2246	0.3860	0.4050	0.1220	0.2720
	0.4082	0.1432	0.3713	0.3005	0.4713
−0.89955	0.3672	0.1892	0.3913	0.0640	0.5659
−0.41695	0.1720	0.4021	0.2732	0.3801	−0.1794
−0.89725		0.30721	0.20070	−0.04334	0.01268
−0.41824	0.4082	0.1432	0.3713	0.3005	0.4713
−0.17751		0.06078	0.05730	−0.20631	0.14194
		0.34239	0.11839	0.25481	−0.37591
			0.09447	−0.27197	0.34997
	0.4082		0.32178	0.19395	0.62848
			0.03628	−0.25154	0.20867
		0.34239	0.11839	0.25481	−0.37591

The largest pivot from unused equations, 0.09447, has just been selected in column x_3.

3.2.2 Multiple RHS and the inverse

If a set of linear equations with alternative RHS vectors is to be solved, the solution process can be carried through with all the RHS columns carried simultaneously. The operations performed are exactly those described in section 3.2.1.

An alternative approach is to develop the inverse of the coefficient matrix.

The inverse of a coefficient matrix A, usually written A^{-1}, is that matrix which when (pre- or post-) multiplied by A, yields the identity matrix I. In practice, the inverse is obtained by carrying out a Gauss, or Gauss–Jordan, process with a special set of RHS vectors representing the identity matrix. The process is usually modified to the extent of dividing the pivot equations by the pivot value as the pivot is selected. The complete process for a 3×3 matrix is shown in table 3.6, the steps,

Table 3.6. *Obtaining inverse by Gauss–Jordan elimination*

Multipliers							Check sum
0.55147	0.225	0.386	0.405	1.	0.	0.	2.016
	−0.408	0.143	0.370	0.	1.	0.	1.105
0.89951	0.367	0.189	0.391	0.	0.	1.	1.165
	0.	0.46486	0.60904	1.	0.55147	0.	2.62537
0.75397	1.	−0.35049	−0.90686	0.	−2.45098	0.	−2.70833
−0.68328	0.	0.31763	−0.05818	0.	0.89951	1.	2.15896
2.76219	0.	1.	1.31016	2.15118	1.18631	0.	5.64766
−0.94379	1.	0.	−0.44766	0.75397	−2.03519	0.	−0.72888
	0.	0.	−0.47432	−0.68328	0.52270	1.	0.36510
	0.	1.	0.	0.26383	2.63011	2.76219	6.65614
	1.	0.	0.	1.39884	−2.52851	−0.94379	−1.07346
	0.	0.	1.	1.44055	−1.10200	−2.10828	−0.76973
	1.	0.	0.	1.39884	−2.52851	−0.94379	
	0.	1.	0.	0.26383	2.63011	2.76219	
	0.	0.	1.	1.44055	−1.10200	−2.10828	

except the final step, being analogous to those set out in table 3.5. The final step consists in rearranging the rows of the fourth block until the nonzero elements of the first three columns lie on the diagonal of the matrix formed by the first three rows and columns, that is, until they form an identity matrix. The inverse matrix is then formed by the last three rows and columns, as thus rearranged, and is indicated by the box in table 3.6. Once an inverse has been obtained, it may be used to obtain solutions for the original equations with any number of RHS vectors. In algebraic form the equation $Ax = b$ has solution $x = A^{-1}b$ which, given A^{-1}, is a multiplication process.

If there are a number of RHS vectors, they may be represented by a matrix B (not necessarily square) in which case the solutions are represented by a matrix X where

$$X = A^{-1}B$$

which is again a multiplication process.

It is never advantageous to develop the inverse matrix if one is only intending to solve sets of linear equations using it. Current practice is to develop matrix equivalents of the Gauss elimination process and to apply these to one or many RHS vectors. This is particularly true if the equations are sparse, i.e. if most coefficients in A are zero. The matrix equivalents are the so-called LU factors of a matrix A, where

$$A = LU$$

with L being lower triangular and U being upper triangular. An example is

$$\underbrace{\begin{bmatrix} 1 & -6 \\ 7 & -37 \end{bmatrix}}_{A} = \underbrace{\begin{bmatrix} 1 & 0 \\ 7 & 5 \end{bmatrix}}_{L} \underbrace{\begin{bmatrix} 1 & -6 \\ 0 & 1 \end{bmatrix}}_{U}$$

The solution procedure for the set of linear equations

$$Ax = LUx = b$$

proceeds in two steps. First define $Ux = y$ and solve

$$Ly = b$$

for y. Then solve

$$Ux = y$$

for x. The first step is a forward substitution corresponding exactly to the forward elimination step of a Gaussian elimination. The second is a backward substitution step and is exactly the backward substitution of the Gaussian elimination method. To illustrate this we solve

$$x_1 - 6x_2 = 5$$
$$7x_1 - 37x_2 = 0$$
$$\begin{bmatrix} 1 & -6 \\ 7 & -37 \end{bmatrix} \begin{bmatrix} x_1 \\ x_2 \end{bmatrix} = \begin{bmatrix} 5 \\ 0 \end{bmatrix}$$

where the coefficient matrix is the one for which we just showed the LU factors. Then

$$Ly = \begin{bmatrix} 1 & 0 \\ 7 & 5 \end{bmatrix} \begin{bmatrix} y_1 \\ y_2 \end{bmatrix} = \begin{bmatrix} 5 \\ 0 \end{bmatrix}$$

or

$$y_1 = 5$$

$$y_2 = \tfrac{1}{5}(0 - 7(5)) = -7$$

(a simple forward substitution). Next

$$Ux = \begin{bmatrix} 1 & -6 \\ 0 & 1 \end{bmatrix} \begin{bmatrix} x_1 \\ x_2 \end{bmatrix} = y = \begin{bmatrix} 5 \\ -7 \end{bmatrix}$$

or

$$x_2 = -7$$

$$x_1 = 5 + 6(-7) = -37$$

(a simple backward substitution).

Finding the factors L and U is relatively straightforward and is covered in Duff (1977). We shall develop an equivalent approach to LU factorization in the next section.

3.2.3 Methods for sparse sets of equations

In the equations which represent the material and thermal balances within a large chemical plant the situation is very different to that in the equations that have just been described, in that only a few of the variables are represented in any one equation with nonzero coefficients, and only a few of the equations have nonzero RHS elements. The coefficient matrix is said to be sparse, and preservation of this sparseness is an important consideration in the solution processes adopted. The importance of this may readily be seen. Consider a set of 1000 equations in 1000 unknowns, then the full matrix augmented with the RHS would contain 1 001 000 elements. But if, on average, each equation contained only five nonzero elements, and if there were 50 nonzero RHS elements, then the number of nonzero elements would be only 5050, a mere 0.5% of the total number of elements.

There is no proven method for solving such sparse linear equations while retaining the greatest possible sparseness. A very extensive survey of the sparse matrix literature is contained in Duff (1977). The method recommended here has been shown to be efficient in many cases, but is not always optimal. It has, however, been well tried in the process simulation field and may be recommended confidently. A full description follows. It is based on the work of Bending and Hutchison (1973).

Given a large sparse set of linear equations, in which only the nonzero coefficients are stored, then a variable and an equation may be described as 'used' when a pivot element relating them has been selected. We then define the 'rank' of an unused equation as the number of nonzero elements it has in as yet unused variables, and the 'file' of an unused variable as the number of nonzero elements it has in as yet unused equations. Using these definitions, it is possible to define a selection procedure for the pivots in what is otherwise a Gaussian elimination procedure quite closely similar to that defined in section 3.2.1.

3. (*a*) Select the variable of least file; if there are several variables of least file, choose arbitrarily
    (*b*) Of the equations referenced by that variable, select the equation of least rank; if there are several equations of least rank, choose the largest element
    (*c*) Check that the element so selected is greater than the user-selected minimum acceptable value, if not, set this variable on one side temporarily and go back to 3(*a*)
    (*d*) Perform the usual operations of Gauss elimination using the selected pivot, and return to 3(*a*)

The operation of this scheme is set out in tables 3.7 and 3.8 for a simple set of flow dividers with recycle. It is left for the reader to reconstruct the system, as an exercise, from the information in table 3.7(*a*).

From table 3.7(*a*) it is seen that columns 2 and 8 contain only one element each. These elements must therefore be selected as pivots for equations 1 and 8 respectively. Since there are no other elements in these columns, no elimination steps are necessary. For the next stage, columns 3, 5, 7 and 9 each contain only two elements. Arbitrarily selecting column 3, there is a choice of pivots between that in equation 2 and that in equation 9, of which that in equation 2 is preferred because of the lower rank of equation 2. The element in equation 9 must be eliminated and in so doing a new element must be created in column 1, equation 9.

This process, which continues in an obvious fashion is set out in a straightforward code in table 3.8.

In practice, of course, the elements of a sparse matrix have to be held and manipulated in some condensed form. One such method will be discussed in the next section.

The scheme just outlined can be applied to situations in which the number of equations supplied is not equal to the number of variables. If too many equations are provided, the algorithm stops when a pivot has been selected for each variable. For certain equations, apparently arbitrarily chosen, no pivot is found, and such equations are not used in the solution process. If insufficient equations are provided, the algorithm will be unable to find a suitable pivot for some of the variables, and these variables cannot be evaluated. However, it will

Table 3.7

Equation no.	1	2	3	4	5	6	7	8	9	RHS
	\[Variable no.\]									
(a) Original equations										
1	−0.333	1								
2	−0.667		1							
3	1				−1				−1	
4				−0.333	1					
5				−0.667		1				
6						−0.333	1			
7								1		1
8						−0.667		1		
9			−1	1			−1			
(b) Equations after pivot selections as in table 3.7(*a*)										
1	−0.333	<u>1</u>								
2	−0.667		<u>1</u>							
3	1			−0.333						1
4				−0.333	<u>1</u>					
5				−0.667		1				
6						−0.333	<u>1</u>			
7									<u>1</u>	1
8						−0.667		<u>1</u>		
9	1.666			1		−0.333				
(c) Equations after complete pivot selection										
1	−0.333	<u>1</u>								
2	−0.667		<u>1</u>							
3	<u>1</u>			−0.333						1
4				−0.333	<u>1</u>					
5				−0.667		<u>1</u>				
6						−0.333	<u>1</u>			
7									<u>1</u>	1
8						−0.667		<u>1</u>		
9				<u>0.555</u>						0.667
(d) Results after backsubstitution step										
	1.400	0.46	0.93	1.200	0.400	0.800	0.26	0.53	1.000	

often occur in sparse systems that much of the backsubstitution can be successfully carried through, even if there are some variables which cannot be evaluated. Some, at least, and perhaps all but a few, of the variables can be correctly evaluated.

Finally, it may be that some of the equations are redundant, in that each contains exactly the same information as some other equation or linear combination of equations. In this case the redundant equations disappear, the coefficients all become exactly zero, or very small numbers comparable with rounding errors. The algorithm is then unable

Table 3.8. *Selection of pivots for a sparse matrix*

(*a*) Select V2E1
     Select V8E8
     Select V3E2   Eliminate V3E9   Create V1E9
     Select V7E6   Eliminate V7E9   Create V6E9
     Select V5E4   Eliminate V5E3   Create V4E3
     Select V9E7   Eliminate V9E3   Create RHSE3

     The results of the process to this point are shown in table 3.7(*b*).

(*b*)
     Select V1E3   Eliminate V1E9   Alter V4E9
                                                Create RHSE9
     Select V6E5   Eliminate V6E9   Alter V4E9
     Select V4E9 as the remaining pivot

     The results of the process to this point are shown in table 3.7(*c*).

     Backsubstitute, using the selected pivots in the reverse order, to obtain the results shown in table 3.7(*d*).

to find a suitable pivot for one or more of the variables, so that these variables cannot be evaluated. But, as in the case of insufficient equations, it may be possible to obtain a partial solution. The more mathematical reader may note that in this situation the matrix of coefficients is singular, but that this does not necessarily preclude a meaningful partial solution to the problem.

3.2.4 Storage and repeated solution of sparse sets of equations

In this section is described one method of storing sparse sets of equations, and the related method of achieving repeated solution of a set of equations. The importance of repeated solution lies in the fact that the method is applicable where one or more of the nonzero elements of the set of equations changes in value (for example, due to an iterative procedure) between applications of the solution process.

In the method to be described, the information about the nonzero elements is held in three lists into which are entered, in arbitrary order, each element in turn. The first and second lists are integer lists and contain respectively the variable number and the equation number of the element. If the element is an RHS element, then the variable number 'zero' is assigned. The third list contains the value of the element. Thus the serial number of the element gives access to all the information about that element required for the solution process. To illustrate this, in table 3.9, arbitrary serial numbers have been assigned to the nonzero elements occurring in table 3.7(*a*). Element 14, for example, relates variable 4 to equation 5 and has value -0.667. In table

Table 3.9. *Arbitrary assignment of serial numbers to elements of table 3.7(a)*

Variable number	1	2	3	4	5	6	7	8	9	RHS
Equation number										
1	2	12								
2	4		3							
3	11				15				20	
4				13	16					
5				14		17				
6						18	8			
7									6	5
8						9		19		
9			10	1			7			

$3.10(a)$ all the 20 nonzero elements of table $3.7(a)$ are listed in order of the arbitrary serial numbers assigned in table 3.9. The solution process of table 3.7 can be applied to the equations sorted in list form, but it is instructive to apply the algorithm of section 3.2.3 directly to table $3.10(a)$. Table $3.10(b)$ shows the created and altered elements that are generated in the solution process.

Repeated solution becomes possible because an element can be completely described by its serial number. The process of creating a new element or altering an existing one can therefore be described in a simple code using the serial numbers. For example, the algebraic representation of element 21, a_{21}, is

$$a_{21} = 0 - a_{10} \cdot a_4/a_3$$

and of element 25 is

$$a_{25} = a_1 - a_{21} \cdot a_{23}/a_{11}$$

The process represented by these algebraic relationships can therefore be represented by the code strings 21, 0, 10, 4, 3 and 25, 1, 21, 23, 11 respectively, whatever the actual values of the elements described, provided only that neither a_3 nor a_{11} is, or becomes as a result of iteration, zero or very small. It is easily seen that the whole solution process may be described by a succession of five-number codes; in practice it is possible to condense the code strings somewhat, as can be seen by considering the elimination of several elements from the same column. The complete succession of code strings may be called an operator list, because it contains all the information necessary to operate on the value list of the original equations in order to achieve a solution. It may also be applied to the value list of any similar set of equations, similar in the sense of having nonzero coefficients wherever the original equations have nonzero coefficients.

Construction of the operator list requires fairly extensive searching of

Table 3.10

Serial no.	Variable no.	Equation no.	Original value	
(a) Initial lists				
1	4	9	1.0	**25
2	1	1	−0.333	
3	3	2	1.0	P3
4	1	2	−0.667	
5	0	7	1.0	
6	9	7	1.0	P5
7	7	9	−1.0	*
8	7	6	1.0	P4
9	6	8	−0.667	
10	3	9	−1.0	*
11	1	3	1.0	P7
12	2	1	1.0	P1
13	4	4	−0.333	
14	4	5	−0.667	
15	5	3	−1.0	*
16	5	4	1.0	P5
17	6	5	1.0	P8
18	6	6	−0.333	
19	8	8	1.0	P2
20	9	3	−1.0	*
(b) Subsequently created elements and alterations (which normally overwrite original values)				
21	1	9	−0.667	
22	6	9	−0.333	
23	4	3	−0.333	
24	0	3	1.0	
25	4	9	0.777	**27
26	0	9	0.667	
27	4	9	0.555	P9

*, Deleted element.
**n, Altered element, stored in serial number n.
Pm, mth selected pivot element.

the variable number, equation number and value lists, and is therefore a relatively slow process. Use of the operator list, by contrast, does not require the use of the variable number or equation number lists, but only direct extraction and manipulation of elements from the value list, and is very rapid. Such a process becomes progressively more advantageous as the size of the problem is increased.

In many respects, the operator list replaces the inverse of a matrix as a tool for obtaining solutions to problems with multiple RHS vectors, either known at one time or developed sequentially. Repeated solution using an operator list will provide a fast means of generating solutions for cases with multiple or sequentially developed RHS vectors as special cases of the situation where any of the elements of the matrix may be changed.

Another generally available tool is the sparse matrix package MA28, available in the Harwell subroutine library (1973). MA28 is a system of Fortran routines which can be used to solve sets of sparse linear equations, and should certainly be investigated by anyone interested in this task.

3.3 General approaches to solving sets of nonlinear equations

We shall consider in this section methods which have proved useful for solving the large sets of nonlinear algebraic equations arising in flow-sheeting problems. The first section considers cases where the equations are available explicitly; the second and third sections consider cases where they are not.

3.3.1 Direct linearization methods

One method of solving a set of nonlinear equations is by direct linearization, that is, by representing the complete set of nonlinear equations by an equivalent set of linear equations, and solving these by the methods of section 3.2. The representation involves approximation and this gives rise to an iterative process in which the linear equations become a better and better approximation to the nonlinear equations as the solution is approached. The equivalent set of linear equations must be so chosen as to yield the same result, of course, as the nonlinear equations. They must also give rise to a convergent iterative process from the chosen starting values to the required solution. It is not always possible to achieve this; on the other hand, it may be achievable in more than one way, and if so, the different rates of convergence to the required solution will be of interest.

As a simple example, consider the dissociation, in dilute solution and at constant temperature, of a species A to two molecules of species B.

$$A \rightleftharpoons 2B$$

Then the following equations have to be solved, where the first equation represents conservation of mass, and the second, thermodynamic equilibrium.

$$C_A + \tfrac{1}{2}C_B = C_A^* \tag{3.19}$$

$$KC_A - C_B^2 = 0$$

C_A and C_B are the concentrations of species A and B respectively, C_A^* is the initial concentration of C_A, and K is the equilibrium constant for the reaction.

Let us assume an approximate solution $C_B^{(p)}$ for the concentration of species B, where $C_B^{(p)}$ is the value for C_B after the pth iteration. One method of linearization would be to write the equations in the form

$$\begin{bmatrix} 1 & \tfrac{1}{2} \\ K & -C_B^{(p)} \end{bmatrix} \begin{bmatrix} C_A \\ C_B \end{bmatrix} = \begin{bmatrix} C_A^* \\ 0 \end{bmatrix} \tag{3.20}$$

It can easily be verified by algebraic solution of these equations (e.g. by Cramer's rule) that this is equivalent to an iterative scheme.

$$C_B^{(p+1)} = \frac{KC_A^*}{\tfrac{1}{2}K - C_B^{(p)}} \tag{3.21}$$

Alternatively, the equations can be linearized in the form

$$\begin{bmatrix} 1 & \tfrac{1}{2} \\ K & -2C_B^{(p)} \end{bmatrix} \begin{bmatrix} C_A \\ C_B \end{bmatrix} = \begin{bmatrix} C_A^* \\ -C_B^{(p)2} \end{bmatrix} \tag{3.22}$$

which gives rise to the iterative scheme

$$C_B^{(p+1)} = \frac{KC_A^* + C_B^{(p)2}}{\tfrac{1}{2}K + 2C_B^{(p)}} \tag{3.23}$$

Table 3.11 shows the first few steps of these two iterative procedures for a particular case. While both converge to a solution, it is clear that the second method does so much more rapidly than the first. In fact it can be shown that the process described by equation (3.23) is a second-order process, entirely equivalent to a Newton–Raphson-type iteration. A second-order process is one in which the number of significant digits of accuracy tends to double with each iteration as one gets very close to the solution.

There is also another solution to the equations (3.19), at $C_B = -2.0$, but it is not a physically meaningful solution. However, the two processes have quite different convergence properties in the neighbourhood of the negative solution. Here the second-order process (equation (3.23)) converges to the genuine (but not physically meaningful) solution, while the first-order process (equation (3.21)) diverges from the negative solution and ends up with the (physically meaningful) solution, $C_B = 1.0$. The divergence from the negative solution is, however, very slow in the near neighbourhood of that solution, so that special tests might have to be devised to test for it. With this proviso, it may be shown that the process of equation (3.21) will converge to the physical solution from any starting value $-\infty < C_B^{(0)} < +\infty$, while $C_B^{(0)} \neq -2.0$; the second-order process of equation (3.23) will converge to the physical solution from any starting point $-0.5 < C_B^{(0)} < +\infty$, and to the alternative solution from any starting point $-\infty < C_B^{(0)} < -0.5$. The convergence from $C_B^{(0)} = \pm n$ for the second-order method, where n is any large number is, however, slow and of first order. In physical reality, $C_B^{(0)}$ is bounded, $0 < C_B^{(0)} < 2$, and from any point within that range convergence is assured for either method, but more rapidly for the method of equations (3.22) or (3.23).

Table 3.11. *Comparison of two iterative solutions to linearized equations*

$$C_B^{(p+1)} = \frac{KC_A^*}{\frac{1}{2}K + C_B^{(p)}} \qquad\qquad C_B^{(p+1)} = \frac{KC_A^* + (C_B^{(p)})^2}{\frac{1}{2}K + 2C_B^{(p)}}$$

For $K = 2$, $C_A^* = 1$, and $C_B^{(0)} = 1.5$

$C_B^{(0)} = 1.5$	$C_B^{(0)} = 1.5$
$C_B^{(1)} = 0.8$	$C_B^{(1)} = 1.0675$
$C_B^{(2)} = 1.111111$	$C_B^{(2)} = 1.001250$
$C_B^{(3)} = 0.947368$	$C_B^{(3)} = 1.000001$
$C_B^{(4)} = 1.027027$	$C_B^{(4)} = 1.000000$
$C_B^{(5)} = 0.986667$	
$C_B^{(6)} = 1.006711$	
$C_B^{(7)} = 0.996656$	
$C_B^{(8)} = 1.001675$	
$C_B^{(9)} = 0.999163$	
$C_B^{(10)} = 1.000417$	

Analytic solution, $C_B = 1.0, -2.0$

It is, clearly, also possible to consider iteration schemes based upon sets of linear equations that are linear combinations of equations (3.20) and (3.22). Such combinations will, of course, always have the same solutions as the equations from which they have been formed. However, it is possible in principle to weight the combination with a parameter which changes from step to step in the iteration, so that the desirable property of convergence from a wide range of starting values is achieved initially, with a gradual change to the rapid convergence of the second-order process as the computation proceeds.

It will be seen, even from this discussion of a very simple case, that it is difficult to generalize about the rate of convergence, or about the region from within which convergence to a desired solution is possible. The general treatment which follows proves only that it is possible to construct a linearization method which, *from a starting point sufficiently close to a suitable solution*, will converge to that solution in a second-order manner.

Suppose a set of n nonlinear equations in the n variables x, such that

$$f(x_1, x_2, \ldots, x_n) = 0$$

of which the kth equation

$$f_k(x_1, x_2, \ldots, x_n) = 0$$

Then the Jacobian row of the kth equation may be written

$$j_k^{\mathrm{T}} = \frac{\partial f_k}{\partial x_1}, \frac{\partial f_k}{\partial x_2}, \ldots, \frac{\partial f_k}{\partial x_n}$$

and the Jacobian matrix J of the functions f is the assembly of the Jacobian rows j_m^{T}, $m = 1, \ldots, n$.

By expanding f as a Taylor series in the derivatives, it is easily shown (Carnahan, Luther and Wilkes, 1969) that a Newton–Raphson-type iterative procedure can be defined as

$$x_{\mathrm{NR}}^{(p+1)} = x^{(p)} - (J^{(p)})^{-1} f^{(p)} \qquad (3.24)$$

where $x^{(p)}$ is the pth estimate of the solution, $(J^{(p)})^{-1}$ is the inverse of the Jacobian matrix, and $f^{(p)}$ the function values, evaluated using the pth estimate of the solution. The subscript 'NR' indicates that this is a Newton–Raphson-type estimate.

By contrast, we may define a quasi-linear, or 'QL', process in the following way. Represent the functions by the linear equations:

$$f(x) \simeq Ax + b = 0 \qquad (3.25)$$

(as in equations (3.20) and (3.22)) where the coefficients in A may themselves be functions of the current solution estimate. Then

$$x_{\mathrm{QL}}^{(p+1)} = -(A^{(p)})^{-1} b^{(p)} \qquad (3.26)$$

Substituting equation (3.25) in equation (3.24)

$$x_{\mathrm{NR}}^{(p+1)} = x^{(p)} - (J^{(p)})^{-1} (A^{(p)} x^{(p)} + b^{(p)}) \qquad (3.27)$$

Whence if, and only if, $J^{(p)} \equiv A^{(p)}$,

$$x_{\mathrm{NR}}^{(p+1)} = x^{(p)} - (A^{(p)})^{-1} (A^{(p)} x^{(p)} + b^{(p)}) = -(A^{(p)})^{-1} b^{(p)} = x_{\mathrm{QL}}^{(p+1)} \qquad (3.28)$$

The condition that $J^{(p)} \equiv A^{(p)}$ implies that, to obtain second-order convergence, it is necessary and sufficient that the coefficients of the linear equations be chosen such that they are the partial derivatives of the nonlinear functions with respect to the appropriate variables. For an application of this principle to multicomponent vapour–liquid equilibrium, see Hutchison and Shewchuk (1974).

In more general terms, if one of the equations has the general form

$$f_k(x_1, x_2, \ldots, x_n) = 0 \qquad (3.29)$$

then the linearized form will be

$$\sum_{i=1}^{n} \left(\frac{\partial f_k}{\partial x_i}\right)^{(p)} x_i = - f_k \left(x_1^{(p)}, x_2^{(p)}, \ldots, x_n^{(p)}\right)$$

$$+ \sum_{i=1}^{n} \left(\frac{\partial f_k}{\partial x_i}\right)^{(p)} x_i^{(p)} \tag{3.30}$$

where the coefficients on the LHS, and all the terms on the RHS, can be evaluated, given $x_i^{(p)}$, the pth approximation to the solution. This is a completely general formulation.

An important simplification arises if, as is often the case in process simulation, the actual subset of the variables x occurring in the equations (3.30) is a set of composition variables of a particular stream. In that case the RHS of equation (3.30) is identically zero and we have

$$\sum_{s} \left(\frac{\partial f_k}{\partial x_s}\right)^{(p)} x_s = 0 \tag{3.31}$$

where s is restricted to those variables forming the set of composition variables. This result is analogous to the Gibbs–Duhem theorem, and applies only when the subset of variables concerned is strictly confined to one (or more) complete set(s) of composition variables.

As an example of this last point, consider the vapour fraction of the first component of a three-component mixture in equilibrium with the liquid mixture:

$$y_1 - f(x_1, x_2, x_3) = 0 \tag{3.32}$$

This equation may be linearized, using the general result, equation (3.30), to

$$y_1 - \left(\frac{\partial f}{\partial x_1}\right)^{(p)} x_1 - \left(\frac{\partial f}{\partial x_2}\right)^{(p)} x_2 - \left(\frac{\partial f}{\partial x_3}\right)^{(p)} x_3$$

$$= -y_1^{(p)} + f(x_1^{(p)}, x_2^{(p)}, x_3^{(p)})$$

$$+ y_1^{(p)} - \left(\frac{\partial f}{\partial x_1}\right)^{(p)} x_1^{(p)} - \left(\frac{\partial f}{\partial x_2}\right)^{(p)} x_2^{(p)} - \left(\frac{\partial f}{\partial x_3}\right)^{(p)} x_3^{(p)}$$

and simplified to

$$y_1 - \left(\frac{\partial f}{\partial x_1}\right)^{(p)} x_1 - \left(\frac{\partial f}{\partial x_2}\right)^{(p)} x_2 - \left(\frac{\partial f}{\partial x_3}\right)^{(p)} x_3 = 0$$

For further discussion of this point of linearization theory, see Hutchison and Shewchuck (1974).

It is, of course, not necessary to achieve the *algebraic* solution of equations (3.20) and (3.22) in the form of equations (3.21) and (3.23). In practice the often large and sparse sets of linearized equations are solved by the methods of section 3.2, the coefficients and RHS elements recomputed using the new solution vector, and the linearized equations resolved, perhaps by the method of section 3.2.3. This process forms the basis of the plant-simulation method described in chapter 9.

An interesting interpretation of equation (3.24)

$$x^{(p+1)}_{\text{NR}} = x^{(p)} - (J^{(p)})^{-1} f^{(p)} \tag{3.24}$$

is that the last term incorporates both a direction and a step length in which to move from the existing position p, to the new position, $p+1$. Any process which involves a matrix not identical to the Jacobian matrix, say

$$x^{(p+1)}_{Q} = x^{(p)} - (Q^{(p)})^{-1} f^{(p)} \tag{3.33}$$

may be interpreted as a step in a different direction and with a different step length, so that the processes embodied in equations (3.20) and (3.22) may be regarded as two possible steps, in different directions, 'towards' the solution.

Clearly, it is a simple matter to incorporate into equation (3.33) a parameter that would offer some control over the step length, normally to decrease it, but perhaps occasionally to increase it. And, further, it is a simple matter to blend two iterative processes in such a manner as to lead to a step in a direction between that of either of the processes separately.

There is an iteration process due to Marquardt which can be interpreted in this fashion (Marquardt, 1963). Marquardt's analysis leads one to consider two directions: one is the Newton–Raphson direction, as given by equation (3.24); the other is given by the direction of steepest descent of an auxiliary function g_p, which is defined as the sum of squares of the functions f_p. The iteration step is defined by

$$((J^{(p)})^{\text{T}} J^{(p)} + \lambda I)\,(x^{(p+1)} - x^{(p)}) = -(J^{(p)})^{\text{T}} f^{(p)} \tag{3.34}$$

where λ is a single parameter, to be separately determined, which may be varied from 0 to $+\infty$. When $\lambda = 0$, this reduces to the Newton–Raphson process. When λ is large this reduces to a small step in the direction of steepest descent of the auxiliary function, and for intermediate values an intermediate step is obtained, see figure 3.8 for a two-dimensional illustration.

From the point of view of process simulation, a difficulty would appear to be that the matrix product $J^{\text{T}} J$ may be considerably less sparse than the matrix J, so that the work involved in solving equation (3.34) may be rather greater than for the simple iteration process.

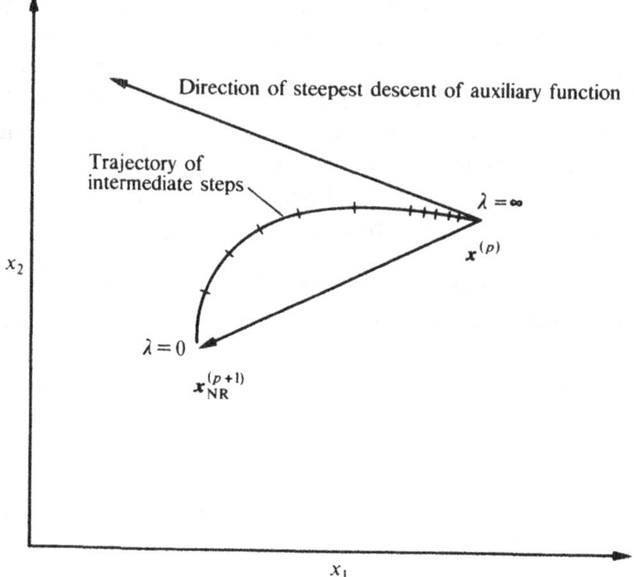

Figure 3.8. Marquardt's process illustrated.

3.3.2 Indirect methods

Often one does not have the functions f available in explicit form, or one may be unwilling to develop the needed partial derivatives algebraically. Figure 3.9 indicates the nature of the general iteration scheme where only the functions f can be evaluated. A common example is if we are supplied with a computer subroutine which we then use to evaluate the functions f given values for the variables x. The functions may have two forms: (1) they may provide calculated values for the variables, as happens for the recycle variables in a stream for the sequential modular approach, or (2) they may only provide errors which should be zero when correct variable values are found, as happened in our isothermal-

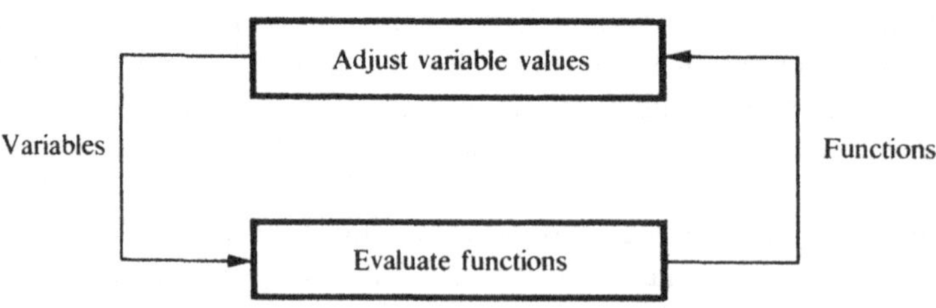

Figure 3.9. Indirect methods.

flash-unit example when we supplied values for V/F and calculated the function $\Sigma x_i - \Sigma y_i$. This difference can be stated as follows.

Case (1) $x^{(\text{calc})} = f(x^{(\text{guess})})$ $(= x^{(\text{guess})}$ if $x^{(\text{guess})}$ is the solution)

Case (2) $e = g(x^{(\text{guess})})$ $(= 0$ if $x^{(\text{guess})}$ is the solution)

Case (1) can always be converted to case (2) by letting

$$e = x^{(\text{calc})} - x^{(\text{guess})}$$

We can label case (1) an 'explicit iteration loop' and case (2) an 'implicit iteration loop'. The explicit loop provides us with more information, and we can exploit this to converge the variable values. A commonly used method to obtain new variable values for explicit loops is called successive (or direct) substitution, where we simply use the newly calculated values as our next guess, i.e.

$$x^{(\text{next})} = x^{(\text{calc})} = f(x^{(\text{guess})}) \tag{3.35}$$

More generally one would use a weighted average of the last guess and the calculated result:

$$x^{(\text{next})} = ax^{(\text{calc})} + (1-a)\,x^{(\text{guess})} \tag{3.36}$$

We might illustrate this approach with a simple example. Suppose we wish to solve

$$x = \cos(x); \; x \text{ in radians}$$

for x, a problem we solved earlier in section 3.1. An explicit iteration loop we could propose is

$$x^{(\text{calc})} = \cos(x^{(\text{guess})})$$

and we would get the following results using successive substitution with a first guess of $x = 0.5$.

Iteration	$x^{(\text{guess})}$	$\cos(x^{(\text{guess})})$
1	0.5	0.8776
2	0.8776	0.6390
3	0.6390	0.8027
4	0.8027	0.6948
5	0.6948	0.7682
6	0.7682	0.7192
7	0.7192	0.7524
8	0.7524	0.7301
9	0.7301	0.7451
10	0.7451	0.7350
.	.	.
.	.	.
.	.	.
23	0.7391	0.7391

Looking at the numbers we see that they oscillate, so we might choose to use the average of the last guess and the calculated result, i.e. we would be using equation (3.36) with $a = 0.5$.

Iteration	$x^{(guess)}$	$\cos(x^{(guess)})$
1	0.5	0.8776
2	0.6888	0.7720
3	0.7304	0.7449
4	0.7377	0.7400
5	0.7389	0.7392
6	0.7390	0.7391
7	0.7391	0.7391

Convergence has obviously improved quite substantially.

We might wonder if there exists a best value for a, perhaps one which we would recalculate at each step. By remembering our experience with the secant method, we know that we can adopt that approach. We may write an error function

$$e = x^{(calc)} - x^{(guess)} = \cos(x^{(guess)}) - x^{(guess)}$$

and have, except for a factor of -1, exactly the problem we solved earlier with the secant method. The one difference is that we do not need two values of x to start. By using

$$x_1 = x^{(guess)}$$

$$x_2 = \cos(x^{(guess)})$$

we can then use the secant method to find the root. With some algebraic manipulation the reader can discover the value that a should assume in equation (3.36) so that the secant method results. This equation for a appears quite commonly in the literature. Remember that a will vary for each iteration here.

While the above seems fairly straightforward, it is its application to several functions in several unknowns that becomes interesting. The key points we are stressing here are:

(1)   An explicit iteration provides us with a means to perturb the variables automatically before we apply any technique that we might devise to accelerate the convergence.

(2)   Even without any acceleration, the method of successive substitution may converge. We should not overlook that it may also diverge. However, for a number of physical problems, such as those relating to flowsheeting, one can produce a qualitative argument that they ought to converge, although they may do so very slowly.

We shall now look at some particular convergence-acceleration techniques.

The generalized secant method

Assume we have t functions g in t variables x. We shall assume that the problem involves only implicit loops. As we did for the single function, single variable case, we can attempt to solve by finding the linear functions which approximate our nonlinear functions. We can set these to zero and solve for their roots and hope that these will approximate the roots of our nonlinear functions. Suppose we perform the following numerical experiment

$x^{(\text{guess})}$	$g(x^{(\text{guess})})$
$x^{(0)}$	$e^{(0)} = g(x^{(0)})$
$x^{(1)}$	$e^{(1)} = g(x^{(1)})$
$x^{(2)}$	$e^{(2)} = g(x^{(2)})$
.	.
.	.
.	.
$x^{(t)}$	$e^{(t)} = g(x^{(t)})$

where we have evaluated the error function

$$e = g(x^{(\text{guess})})$$

at $t+1$ different values for the variables x. We can write an approximate linear model which will fit these data exactly:

$$e^* = Ax + b \tag{3.37}$$

so

$$e^{(0)} = Ax^{(0)} + b$$

$$e^{(1)} = Ax^{(1)} + b$$

$$\cdot$$

$$\cdot$$

$$\cdot$$

$$e^{(t)} = Ax^{(t)} + b$$

We have $t+1$ sets of t linear equations, and the unknowns are the $t \times t$ elements of A and the t elements of b. Thus we have $t^2 + t$ linear equations in $t^2 + t$ unknowns, and we can solve. The problem can be written compactly as follows. First eliminate b by writing

$$\Delta e^{(1)} = e^{(1)} - e^{(0)} = A(x^{(1)} - x^{(0)}) = A\Delta x^{(1)}$$

$$\Delta e^{(2)} = e^{(2)} - e^{(1)} = A(x^{(2)} - x^{(1)}) = A\Delta x^{(2)}$$

$$\vdots$$

$$\Delta e^{(t)} = e^{(t)} - e^{(t-1)} = A(x^{(t)} - x^{(t-1)}) = A\Delta x^{(t)}$$

or

$$(\Delta e^{(1)}, \Delta e^{(2)}, \ldots, \Delta e^{(t)}) = A(\Delta x^{(1)}, \Delta x^{(2)}, \ldots, \Delta x^{(t)})$$

or

$$\Delta E = A\Delta X$$

Thus

$$A = \Delta E \Delta X^{-1} \tag{3.38}$$

provides us with A if, and only if, ΔX has an inverse. Since ΔX is completely under our control (we choose how to perturb x), we can guarantee an inverse if we run into difficulty. The value for b can be found from

$$b = e^{(t)} - Ax^{(t)} \tag{3.39}$$

We can now estimate the value of $x^{(\text{next})}$ as the one which will make $e^* = 0$. This step involves solving the set of linear equations

$$e^* = 0 = Ax^{(\text{next})} + b$$

or

$$x^{(\text{next})} = -A^{-1}b = -A^{-1}(e^{(t)} - Ax^{(t)})$$

$$= x^{(t)} - A^{-1}e^{(t)} \tag{3.40}$$

The algorithm which is being suggested here is as follows.

1. Find $e = g(x)$ for $t+1$ different values of x
2. Use equations (3.38) and (3.40) to estimate a new x
3. Find $e = g(x)$ for the x estimated in step 2
4. Check if problem has converged (are all the errors in e small enough and/or did the Δx values become too small to continue?)
    (*a*) If not converged, replace one of the $(x^{(i)}, e^{(i)})$ sets used in step 2 (perhaps the oldest) by the new set and iterate from step 2
    (*b*) Otherwise continue
5. Exit

A simple example might help to demonstrate this algorithm. We shall set up the solution to

$$e_1 = 20x_1^{1.1} + x_2 - 30 = 0$$

$$e_2 = -5x_1 + 50x_2^{0.9} - 60 = 0$$

1. Choose $x^{(0)} = \begin{bmatrix} x_1 \\ x_2 \end{bmatrix}^{(0)} = \begin{bmatrix} 0 \\ 0 \end{bmatrix}$, $x^{(1)} = \begin{bmatrix} 1 \\ 0 \end{bmatrix}$ and $x^{(2)} = \begin{bmatrix} 0 \\ 1 \end{bmatrix}$

as our first three points. Then

$$e^{(0)} = \begin{bmatrix} e_1 \\ e_2 \end{bmatrix}^{(0)} = \begin{bmatrix} 20(0)^{1.1} + 0 - 30 \\ -5(0) + 50(0)^{0.9} - 60 \end{bmatrix} = \begin{bmatrix} -30 \\ -60 \end{bmatrix}$$

Similarly

$$e^{(1)} = \begin{bmatrix} -10 \\ -65 \end{bmatrix}, \quad e^{(2)} = \begin{bmatrix} -29 \\ -10 \end{bmatrix}$$

We can conveniently collect these results into two matrices:

$$X = \begin{bmatrix} 0 & 1 & 0 \\ 0 & 0 & 1 \end{bmatrix}, \quad E = \begin{bmatrix} -30 & -10 & -29 \\ -60 & -65 & -10 \end{bmatrix}$$

2. Evaluating A using equation (3.38)

$$A = \Delta E \, \Delta X^{-1} \tag{3.38}$$

$$\Delta E = \begin{bmatrix} -10-(-30) & -29-(-10) \\ -65-(-60) & -10-(-65) \end{bmatrix} = \begin{bmatrix} 20 & -19 \\ -5 & 55 \end{bmatrix}$$

$$\Delta X = \begin{bmatrix} 1-0 & 0-1 \\ 0-0 & 1-0 \end{bmatrix} = \begin{bmatrix} 1 & -1 \\ 0 & 1 \end{bmatrix}$$

So

$$A = \begin{bmatrix} 20 & -19 \\ -5 & 55 \end{bmatrix} \begin{bmatrix} 1 & -1 \\ 0 & 1 \end{bmatrix}^{-1} = \begin{bmatrix} 20 & -19 \\ -5 & 55 \end{bmatrix} \begin{bmatrix} 1 & 1 \\ 0 & 1 \end{bmatrix}$$

$$= \begin{bmatrix} 20 & 1 \\ -5 & 50 \end{bmatrix}$$

Now evaluate the next x using equation (3.40)

$$x^{(\text{next})} = x^{(2)} - A^{-1} e^{(2)} = \begin{bmatrix} 0 \\ 1 \end{bmatrix} - \begin{bmatrix} 20 & 1 \\ -5 & 50 \end{bmatrix}^{-1} \begin{bmatrix} -29 \\ -10 \end{bmatrix}$$

$$= \begin{bmatrix} 0 \\ 1 \end{bmatrix} - \begin{bmatrix} -1.4328 \\ -0.3433 \end{bmatrix} = \begin{bmatrix} 1.4328 \\ 1.3433 \end{bmatrix}$$

3. Evaluate $e = g(x^{(\text{next})})$

$$\begin{bmatrix} 20(1.4328)^{1.1} + 1.3433 - 30 \\ -5(1.4328) + 50(1.3433)^{0.9} - 60 \end{bmatrix} \qquad \begin{bmatrix} 1.0486 \\ -1.9532 \end{bmatrix}$$

Clearly the problem has not converged as both errors, though much smaller, are still too large. We replace in X and E the oldest point ($x^{(0)}$ and $e^{(0)}$) with this newly calculated x and e point. We then repeat from step 2 of the algorithm

The effort involved should now be more apparent. The initialization step implies that the functions must be evaluated $t+1$ times. If we are solving a flowsheet, this means that we have to solve the flowsheet $t+1$ times, which is a nontrivial task. It is of interest obviously to try to use these first $t+1$ evaluations as effectively as possible. We might wonder if we could not, in fact, get the problem to converge towards the answer rather than have to wait until we can apply the above generalized secant step.

From our earlier discussion on explicit iteration loops, we know convergence can sometimes occur without acceleration by using successive substitution. Hopefully the variables will, by themselves, be converging to the answer. For implicit iteration loops we must use other methods to obtain the perturbations. For our example problem with two functions, we simply supplied the points $x^{(0)}$, $x^{(1)}$ and $x^{(2)}$. Ideally we would like the point $x^{(1)}$ to improve e and similarly $x^{(2)}$ to reduce it even further. The following ideas often permit us to move toward the answer during the first $t+1$ steps. They are also computationally much more efficient for large problems.

Update formulas for generalized-secant-based methods

The approach as developed above for the generalized secant method implies that, at each step, we recalculate the A matrix from original data. Equation (3.38), repeated here,

$$A = \Delta E \Delta X^{-1}$$

implies that we perform the following steps:

1. Obtain a new Δe for a new Δx

2.   Replace a column of ΔE and ΔX with this new point
3.   Recompute A

In fact, it is possible to develop algorithms which update A in a quite different manner. We shall present first a generalized secant method. The purpose in presenting it is that we already know one approach for the generalized secant method and 'here is another'. But, more importantly, it leads naturally to Broyden's method, which is widely used with considerable success. This presentation will hopefully expose a strong relationship between the two methods, as only one step in a computational algorithm distinguishes them.

Let us assume we have an estimate for A, call it $A^{(1)}$. Let us also assume we have just evaluated $\Delta e^{(1)}$ for a specified $\Delta x^{(1)}$. We would like a matrix $A^{(2)}$ such that

$$\Delta e^{(1)} = A^{(2)}\Delta x^{(1)} \tag{3.41}$$

Using what is termed a 'rank 1' update, we shall assume we can modify $A^{(1)}$ as follows to obtain $A^{(2)}$:

$$A^{(2)} = A^{(1)} + u^{(1)}(v^{(1)})^{T} \tag{3.42}$$

Note that the matrix $u^{(1)}(v^{(1)})^{T}$ is an 'outer product' of the vectors $u^{(1)}$ and $v^{(1)}$,

$$u^{(1)}(v^{(1)})^{T} = \begin{bmatrix} u_1 v_1 & u_1 v_2 & \cdots & u_1 v_t \\ u_2 v_1 & u_2 v_2 & \cdots & u_2 v_t \\ \cdot & & & \\ \cdot & & & \\ \cdot & & & \\ u_t v_1 & u_t v_2 & \cdots & u_t v_t \end{bmatrix}^{(1)}$$

and this matrix has a rank of 1 only, as each column is a scalar multiple of every other one. Substituting equation (3.42) into equation (3.41) we get

$$\Delta e^{(1)} = [A^{(1)} + u^{(1)}(v^{(1)})^{T}]\,\Delta x^{(1)} \tag{3.43}$$

At this point we shall simply choose $v^{(1)}$ arbitrarily and discover if a formula for $u^{(1)}$ results, which it does:

$$u^{(1)} = (\Delta e^{(1)} - A^{(1)}\Delta x^{(1)})/[(v^{(1)})^{T}\Delta x^{(1)}] \tag{3.44}$$

As long as $v^{(1)}$ is not orthogonal to $\Delta x^{(1)}$ (i.e. the denominator $(v^{(1)})^{T}\Delta x^{(1)}$ is not zero), we have a well-defined $u^{(1)}$ vector. Thus we can indeed convert any arbitrary $t \times t$ matrix $A^{(1)}$ to satisfy equation (3.41) by using a 'rank 1' update. Note that if $\Delta e^{(1)}$ were equal to $A^{(1)}\Delta x^{(1)}$ (i.e. the linearized model correctly estimates $\Delta e^{(1)}$) then $u^{(1)}$ is the zero vector 0, and no update results.

Let us now obtain a second datum point, $\Delta e^{(2)}$ for a step $\Delta x^{(2)}$. ($x^{(2)}$ can be the result of solving equation (3.40): $x^{(2)} = x^{(1)} - (A^{(2)})^{-1}e^{(1)}$. Then $\Delta x^{(2)} = x^{(2)} - x^{(1)}$.) We can then require that

$$\Delta e^{(2)} = A^{(3)}\Delta x^{(2)} = [A^{(2)} + u^{(2)}(v^{(2)})^{\mathrm{T}}]\Delta x^{(2)} \tag{3.45}$$

We can easily find $u^{(2)}$ and $v^{(2)}$ as before, but now we would also like to require $A^{(3)}$ to satisfy our first datum point, namely

$$\Delta e^{(1)} = A^{(3)}\Delta x^{(1)} = [A^{(2)} + u^{(2)}(v^{(2)})^{\mathrm{T}}]\Delta x^{(1)} \tag{3.46}$$

If the term $u^{(2)}(v^{(2)})^{\mathrm{T}}\Delta x^{(1)}$ were the zero vector 0, then equation (3.46) would reduce to equation (3.41) and thus be seen to be true. Selecting $v^{(2)}$ orthogonal to $\Delta x^{(1)}$ will give us the desired result. A candidate is to select $v^{(2)}$ as the vector $\Delta x^{(2)}$ orthogonalized with respect to $\Delta x^{(1)}$. Figure 3.10 shows the orthogonalization indicated. The equation for obtaining $v^{(2)}$ is

$$v^{(2)} = \Delta x^{(2)} - [(\Delta x^{(1)})^{\mathrm{T}}\Delta x^{(2)}]\Delta x^{(1)}/[(\Delta x^{(1)})^{\mathrm{T}}\Delta x^{(1)}] \tag{3.47}$$

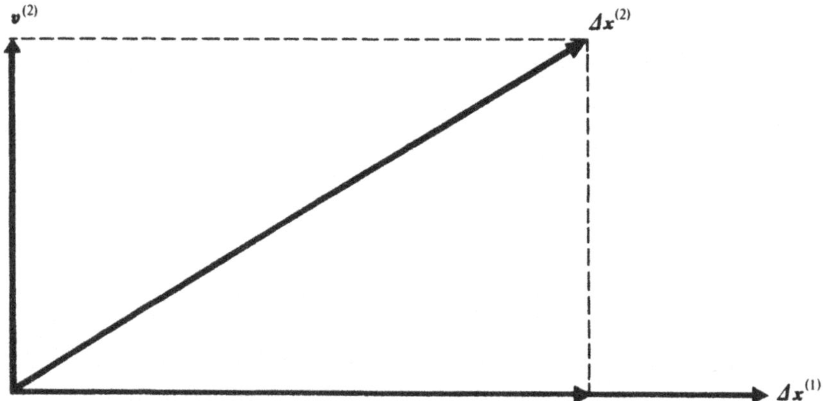

Figure 3.10. $v^{(2)}$ is the component of $\Delta x^{(2)}$ which is orthogonal to $\Delta x^{(1)}$.

We obtain another datum point, $\Delta e^{(3)}$ versus $\Delta x^{(3)}$, and again require

$$\Delta e^{(3)} = A^{(4)}\Delta x^{(3)}$$

along with

$$\Delta e^{(1)} = A^{(4)}\Delta x^{(1)}$$

$$\Delta e^{(2)} = A^{(4)}\Delta x^{(2)}$$

For

$$A^{(4)} = A^{(3)} + u^{(3)}(v^{(3)})^{\mathrm{T}}$$

we find that $v^{(3)}$ should be selected to be orthogonal to both $\Delta x^{(1)}$ and $\Delta x^{(2)}$; this result can be generalized in an obvious way:

$$A^{(i+1)} = A^{(i)} + (\Delta e^{(i)} - A^{(i)}\Delta x^{(i)})(v^{(i)})^{\mathrm{T}}/[(v^{(i)})^{\mathrm{T}}\Delta x^{(i)}] \qquad (3.48)$$

where $v^{(i)}$ is $\Delta x^{(i)}$ orthogonalized with up to $t-1$ previous vectors $\Delta x^{(j)}$.

The algorithm suggesting itself here is as follows:

1. Estimate $A^{(1)}$ (say let $A^{(1)} = I$)
2. Determine $\Delta x^{(1)}$ using equation (3.40), i.e., $x^{(1)} = x^{(0)} - (A^{(1)})^{-1}e^{(0)}$. Then $\Delta x^{(1)} = x^{(1)} - x^{(0)}$. Evaluate resulting $\Delta e^{(1)}(= e^{(1)} - e^{(0)})$
3. Select $v^{(1)} = \Delta x^{(1)}$ and update $A^{(1)}$, getting $A^{(2)}$
4. Determine $\Delta x^{(2)}$ using equation (3.40) and evaluate resulting $\Delta e^{(2)}$
5. Select $v^{(2)}$ equal to the vector component of $\Delta x^{(2)}$ which is orthogonal to $\Delta x^{(1)}$ and update $A^{(2)}$, getting $A^{(3)}$
6. Estimate $\Delta x^{(3)}$ and evaluate resulting $\Delta e^{(3)}$
7. Select $v^{(3)}$ equal to the vector component of $\Delta x^{(3)}$ which is orthogonal to both $\Delta x^{(1)}$ and $\Delta x^{(2)}$; update $A^{(3)}$ getting $A^{(4)}$
8. Etc.

Broyden's method

We shall not fully develop Broyden's method (Broyden, 1965, 1970) but only expose the underlying ideas. Broyden's method is similar to the update method just described, but, as we shall see, it is markedly more efficient in its use of computer space and time. Rather than require that the updated matrix be selected so that the previous steps are satisfied (as for equation (3.46)), Broyden's method requires that

$$A^{(i+1)}z = A^{(i)}z \qquad (3.49)$$

for all vectors z which are orthogonal to $\Delta x^{(i)}$. Updating will require the new A to map the latest $\Delta x^{(i)}$ into the resulting $\Delta e^{(i)}$. All directions *orthogonal* to $\Delta x^{(i)}$ will map through the new A unchanged, whereas, in the generalized secant method, all *previous* $\Delta x^{(j)}$ map through the new A unchanged. The change appears slight, perhaps, but the effect is significant on the computational algorithm. As will be seen the advantage here is that the previous $\Delta x^{(j)}$ steps need not be stored. All the previous history that is kept is just that in A as it is updated. Again using a rank 1 update, we get

$$\Delta e^{(i)} = A^{(i+1)}\Delta x^{(i)} = [A^{(i)} + u^{(i)}(v^{(i)})^{\mathrm{T}}]\Delta x^{(i)}$$

Equation (3.49) also gives

$$[A^{(i)} + u^{(i)}(v^{(i)})^{\mathrm{T}}]z = A^{(i)}z$$

for all vectors $z \neq 0$ which are orthogonal to $\Delta x^{(i)}$. Clearly, if we select $v^{(i)} = \Delta x^{(i)}$, then the inner product $(v^{(i)})^{\mathrm{T}}z = (\Delta x^{(i)})^{\mathrm{T}}z = 0$ if, and only if,

$z \neq 0$ is orthogonal to $\Delta x^{(i)} = 0$. For this choice

$$u^{(i)} = (\Delta e^{(i)} - A\Delta x^{(i)})/[(\Delta x^{(i)})^T \Delta x^{(i)}]$$

and

$$A^{(i+1)} = A^{(i)} + (\Delta e^{(i)} - A\Delta x^{(i)}) \, (\Delta x^{(i)})^T/[(\Delta x^{(i)})^T \Delta x^{(i)}] \qquad (3.50)$$

which is one form of the Broyden update formula.

Given an estimate for $A^{(i+1)}$, the predicted next step $\Delta x^{(i+1)}$ for either the generalized secant method or Broyden's method is given by

$$A^{(i+1)}\Delta x^{(i+1)} = (\Delta e^{(i+1)})_{\text{desired}} = -e^{(i)} \qquad (3.51)$$

Thus we must solve these linear equations to find $\Delta x^{(i+1)}$. Another form of Broyden's formula uses Householder's formula to relate the inverse of $A^{(i+1)}$ to the inverse of $A^{(i)}$ if the two differ by a rank 1 update. Thus this version keeps track of $H^{(i+1)} = (A^{(i+1)})^{-1}$. If $H^{(i+1)}$ is known, then equation (3.51) becomes

$$\Delta x^{(i+1)} = -H^{(i+1)}e^{(i)} \qquad (3.52)$$

which is a simple matrix multiplication to get $\Delta x^{(i+1)}$, eliminating the need to solve the set of linear equations (3.51).

Householder's formula is

$$H^{(i+1)} = (A^{(i+1)})^{-1} = [A^{(i)} + u^{(i)}(v^{(i)})^T]^{-1}$$
$$= (A^{(i)})^{-1} - (A^{(i)})^{-1}u^{(i)}(v^{(i)})^T(A^{(i)})^{-1}/$$
$$[1 + (v^{(i)})^T(A^{(i)})^{-1}u^{(i)}]$$
$$= H^{(i)} - H^{(i)}u^{(i)}(v^{(i)})^T H^{(i)}/[1 + (v^{(i)})^T H^{(i)}u^{(i)}]$$

and the update formula for H becomes

$$H^{(i+1)} = H^{(i)} - (H^{(i)}\Delta e^{(i)} - \Delta x^{(i)}) \, (v^{(i)})^T H^{(i)}/[(v^{(i)})^T H^{(i)}\Delta e^{(i)}] \qquad (3.53)$$

For Broyden's method $v^{(i)}$ is $\Delta x^{(i)}$. For the generalized secant method, it is $\Delta x^{(i)}$ orthogonalized with respect to as many as $t-1$ previous vectors $\Delta x^{(j)}$. The algorithm suggested is (with embellishments)

1. Estimate $H^{(1)} = (A^{(1)})^{-1} = I^{-1} = I$, the identity matrix. User supplies $x^{(0)}$
2. Evaluate the error vector $e^{(0)}$ as a function of $x^{(0)}$. If $(e^{(0)})^T e^{(0)}$ small enough, exit
3. Estimate $\Delta x^{(1)} = -H^{(1)}e^{(0)}$. Let $i = 1$
4. Let $x^{(i)} = x^{(i-1)} + \Delta x^{(i)}$ and evaluate $e^{(i)}$. If $(e^{(i)})^T e^{(i)}$ small enough, exit
5. Evaluate $\Delta e^{(i)} = e^{(i)} - e^{(i-1)}$
6. Evaluate $v^{(i)}$ (as either $\Delta x^{(i)}$ for Broyden's method or as the

orthogonalized component of $\Delta x^{(i)}$ with respect to previous vectors $\Delta x^{(j)}$ for the generalized secant method)

7. Evaluate the denominator of the last term in equation (3.53), $(v^{(i)})^{\mathrm{T}}H^{(i)}\Delta e^{(i)}$. If denominator too small (say less than 10^{-3} times any element in $H^{(i)}$), skip to step 9
8. Update $H^{(i)}$ to get $H^{(i+1)}$ using equation (3.53)
9. Estimate next $\Delta x^{(i+1)} = -H^{(i+1)}e^{(i)}$
10. Set $i = i+1$ and repeat from step 4

The advantage of Broyden's method over a full generalized secant update is, as stated earlier, that no previous vectors $\Delta x^{(j)}$ need be stored. These vectors, for $t = 100$, would require $100 \times 100 = 10\ 000$ storage locations. The disadvantage is that Broyden's method has only super-linear convergence properties (Dennis and More, 1977) whereas the generalized secant method is quadratically convergent. Computational experience seems to show that Broyden's method gives quite favourable convergence in most instances.

One could envisage variations on the above, such as orthogonalizing with only the last k vectors $\Delta x^{(j)}, j < i$, where $k \ll t$. Then only k vectors (say 10) need be stored.

Wegstein acceleration

An acceleration method which is very commonly used in flowsheeting is Wegstein's method (Wegstein, 1958).

Wegstein acceleration is simply to treat each variable by the one-dimensional secant method by driving it with a uniquely associated function; interactions with other variables are ignored. We might try it on our previous generalized secant example. For implicit iteration loops two guesses, $x^{(0)}$ and $x^{(1)}$, are needed to get started; if the loops were explicit we could have started with only $x^{(0)}$ and used the resulting successive substitution values for $x^{(1)}$. We must choose them so that both x values are perturbed.

1. Choose
$$x^{(0)} = \begin{bmatrix} 0 \\ 0 \end{bmatrix}, x^{(1)} = \begin{bmatrix} 1 \\ 1 \end{bmatrix}. \text{ Then}$$
$$e^{(0)} = \begin{bmatrix} -30 \\ -60 \end{bmatrix}, e^{(1)} = \begin{bmatrix} -9 \\ -15 \end{bmatrix}$$

2. To x_1 and x_2 apply the secant method
$$x_1^{(\mathrm{next})} = 1 - \left(\frac{-9-(-30)}{1-0}\right)^{-1}(-9) = 1.4286$$
$$x_2^{(\mathrm{next})} = 1 - \left(\frac{-15-(-60)}{1-0}\right)^{-1}(-15) = 1.3333$$

Therefore

$$x^{(2)} = \begin{bmatrix} 1.4286 \\ 1.3333 \end{bmatrix}$$

and

$$e^{(2)} = \begin{bmatrix} 20(1.4286)^{1.1} + 1.3333 - 30 \\ -5(1.4286) + 50(1.3333)^{0.9} - 60 \end{bmatrix} = \begin{bmatrix} 0.9422 \\ -2.3667 \end{bmatrix}$$

Note that by step 2 we have already significantly reduced the errors in $e^{(2)}$ from those in $e^{(0)}$ and $e^{(1)}$. Note how close $x^{(2)}$ is to $x^{(3)}$ determined

previously, namely $x^{(3)} = \begin{bmatrix} 1.4328 \\ 1.3433 \end{bmatrix}$.

This problem has been specially set up to give rather favourable results. It is almost linear, and the first function is dominated by x_1, and the second by x_2. We might estimate x_1 using the second error function and x_2 using the first, which is the wrong way since x_1 has only a minor effect on e_1 and x_2 a minor effect on e_1. In this case

$$x_1^{(\text{next})} = 1 - \left(\frac{-15 - (-60)}{1-0} \right)^{-1} (-9) = 1.2000$$

$$x_2^{(\text{next})} = 1 - \left(\frac{-9 - (-30)}{1-0} \right)^{-1} (-15) = 1.7143$$

and we get much worse estimates than by doing it the correct way:

$$e^{(2)} = \begin{bmatrix} 20(1.2)^{1.1} + 1.7143 - 30 \\ -5(1.2) + 50(1.7143)^{0.9} - 60 \end{bmatrix} = \begin{bmatrix} -3.8441 \\ 15.2166 \end{bmatrix}$$

The errors are only marginally better than $e^{(1)}$.

We now return to a basic theme again. If each variable strongly influences a particular and unique function, and if we are fortunate enough to select that association, then this type of approach will quite likely enhance the convergence. If we are particularly fortunate, it may even cause convergence in many fewer than t iterations, where t is the number of variables and functions. Thus we might manage to achieve convergence before the first $t+1$ iterations are completed, that is, before we can apply the generalized secant method as initially described.

For explicit iteration loops, we would select the difference in the guessed and calculated value for a variable as the associated error function for that variable. For implicit iteration loops the association of a variable to a particular function may be evident from a physical interpretation of the problem. It is probably best supplied by the person setting up the problem.

In the following discussion we shall look in more depth into what we might term one-dimensional-type acceleration methods, including over-relaxation, the dominant eigenvalue method and, finally, the Wegstein or 'secant method' (which we have just discussed). We shall try to extract from this discussion the reasons why some of the numerical experience often encountered in flowsheeting is as it is when applying these methods.

Overrelaxation

Large linear systems, or linearized systems are often solved by iterative methods to economize on computer memory allocation since most large systems are usually sparse. Iterative methods avoid completely the fill-in that accompanies matrix inversion or related procedures. Certain deductions about linear systems behaviour also apply to the iterative solution of sequential flowsheeting models.

As defined earlier, an approximate linear model of an iterative flowsheeting model is (see equation (3.37))

$$e = Ax + b$$

To discuss the implications of overrelaxation methods, it is convenient to split the A matrix:

$$A = F - I \tag{3.54}$$

where I is an identity matrix. The error vector is now

$$e = Fx - x + b \tag{3.55}$$

In fact, equation (3.55) provides a basis for successive substitution on explicit iteration variables

$$x^{(i+1)} = Fx^{(i)} + b \tag{3.56}$$

provided we interpret the error e as the difference between successive values of x. (The error vector at the ith step is

$$e^{(i)} = x^{(i+1)} - x^{(i)} \tag{3.57}$$

Substituting equation (3.57) into equation (3.55) and superscripting the x vectors in equation (3.55) with (i) yields equation (3.56).)

Since the flowsheeting model is nonlinear, a matrix A may be obtained by fitting, as shown earlier, or by partial differentiation or perturbation. However, it is not necessary to have an explicitly defined matrix. It suffices that given an estimate of the iterate variables $x^{(i)}$ one can generate $x^{(i+1)}$ from the program,

$$x^{(i+1)} = f(x^{(i)}) \tag{3.58}$$

The sequence of x vectors will be considered as having been generated by a locally linearized model like equation (3.56).

The successive substitution procedure, viz. equation (3.58), is analogous in that form to the Jacobi method for solving linear systems. It is well known that the Jacobi procedure will not converge unless the maximum eigenvalue of F is less than 1 in magnitude (Young, 1971). We assume the eigenvalues to be real and positive. It is also a very common experience that mass balance iterations *in sequential modular* flowsheeting converge monotonically. Hence, the implied linear iteration procedure of equation (3.58) does indeed have a maximum eigenvalue less than unity for systems exhibiting this behaviour. The same experience can generally be extended to simultaneous energy and mass balance iterations, but exceptions do arise in the latter case where the enthalpy estimates of a stream may have a drastic effect on the phase condition, introducing discontinuities in the functions f of equation (3.58). We will ignore these exceptional cases (which must be restored to good behaviour either by more careful sequencing of the calculations or by bounding the iterate variables) and apply the ideas to all stream iterate variables.

The rate of convergence of the procedure of equation (3.56) is a function of the largest eigenvalue of F. Young (1971) demonstrates that the number of iterations required to reduce the Euclidian norm of an error vector to a fraction ρ of its initial value is

$$n \geqslant \log \rho / \log M \tag{3.59}$$

where M is the largest eigenvalue of F. Alternatively,

$$\rho = \| e^{(n)} \| \, / \, \| e^{(0)} \| \leqslant M^n \tag{3.60}$$

where the equality would hold for sufficiently large n.

The simultaneous overrelaxation method (JOE) is an attempt to accelerate the iteration procedure by scaling, such that the maximum eigenvalue of the modified procedure (Jacobi iteration plus relaxation) is reduced to a minimal value. The result is accelerated convergence.

Assume that a new iterate vector is obtained by weighting the $(i+1)$th and ith iterates:

$$x^{(i+1)*} = \omega x^{(i+1)} + (1-\omega)x^{(i)} \tag{3.61}$$

i.e.

$$x^{(i+1)*} = \omega(Fx^{(i)} + b) + (1-\omega)x^{(i)} \tag{3.62}$$

Rearranging,

$$x^{(i+1)*} = [\omega F + (1-\omega)I]x^{(i)} + b \tag{3.63}$$

and the eigenvalues of the JOE procedure $\{\lambda_i'\}$ are readily shown to be

$$\lambda_i' = \omega(\lambda_i - 1) + 1 \tag{3.64}$$

where $\{\lambda_i\}$ are the eigenvalues of F.

Remembering that we are assuming all eigenvalues to be real and positive, the optimum choice of ω is the value which scales or maps all of the λ_i into the smallest interval $[M', -M']$. The two most important members of the set $\{\lambda_i\}$ are the largest, M, and smallest, m, since any linear transformation like equation (3.64), which reduces the interval on these, will map the intermediate values between M' and $-M'$. M will be mapped onto M' and m will be mapped onto $-M'$. The obvious solution is to minimize the sum of M' and $-M'$, in fact, to set this sum to zero.

$$M' = \omega(M-1) + 1$$

$$-M' = \omega(m-1) + 1$$

and

$$M' + (-M') = 0 = \omega(M-1) + 1 + \omega(m-1) + 1 \tag{3.65}$$

The optimum relaxation factor ω_0 is then

$$\omega_0 = 2/(2-M-m) \tag{3.66}$$

Values of ω_0 tend to be larger than 1, hence the JOE method extrapolates the Jacobi method. As we will see in later examples, values of ω only slightly greater than ω_0 lead to unstable iteration procedures since the resultant JOE has eigenvalues larger than unity. (We should note that the problem of finding eigenvalues is harder than solving equations so we cannot know M and m by direct methods.)

Acceleration factors for JOE are not absolutely bounded but tend to an upper limit of 2. Optimum improvement in convergence rate then approaches 2, i.e. total iterations to convergence are halved in comparing JOE to successive substitution. This degree of reduction in iteration count is the best that one-dimensional first-order methods like JOE can achieve on multidimensional problems when acceleration is applied at every iteration step.

Example 1. Consider a linear system with two iterate variables whose matrix F has eigenvalues of 0.9 and 0. Using Young's result, the number of iterations to reduce the initial error by a factor of 1000 is, from equation (3.59),

$$n \geq \log \rho / \log M$$
$$\geq \log 0.001 / \log 0.9$$
$$\geq 66$$

Using JOE with
$$\omega_o = 2/(2-0.9-0) = 1.8182$$
the maximum eigenvalue is reduced to
$$\begin{aligned} M' &= \omega_o(M-1)+1 \\ &= 1.8182(0.9-1)+1 \\ &= 0.8182 \end{aligned}$$

Note that no acceleration corresponds to $\omega = 1$. The number of iterations to convergence with JOE acceleration is
$$\begin{aligned} n' &\geq \log 0.001/\log 0.8182 \\ &\geq 35 \end{aligned}$$

To illustrate the sensitivity of this procedure, if an acceleration factor ω_o of 2.0 were used, the maximum eigenvalue of the JOE procedure would correspond to $M = 1.0$ and
$$M' = 2(1.0-1.0)+1.0 = 1.0$$
The procedure would not converge.

Example 2(a). A pathological example is the following linear system:
$$\begin{bmatrix} \hat{x_1} \\ x_2 \end{bmatrix} = \begin{bmatrix} 0.99995 & 0.9999 \\ 0.0000375 & 0.000075 \end{bmatrix} \begin{bmatrix} x_1 \\ x_2 \end{bmatrix} + \begin{bmatrix} 82.0 \\ 0.0 \end{bmatrix}$$

whose eigenvalues are 0.9999874996 and 3.75004×10^{-5}. Thus
$$\omega_o = 2.000050001 \quad \text{and} \quad M' = 0.9999749996$$

Successive substitution would require 552 600 iterations to reduce the initial error by a factor of 1000. The JOE method would require 276 300 iterations. Again JOE improves convergence by a factor of 2 at best.

This example will be taken up again later.

The dominant eigenvalue method

The dominant eigenvalue method (DEM) of Orbach and Crowe (1971) is an interesting member of a class of one-dimensional methods that includes JOE and the secant (or Wegstein) procedure. As the work of Young (1971) has indicated, the rate of convergence of a successive substitution procedure depends on the maximum eigenvalue of F in equation (3.56). Young's formula may be used to estimate the maximum eigenvalues since the ratio of successive error norms after a few iterations should stabilize to a relatively constant value:
$$\|e^{(i+1)}\|/\|e^{(i)}\| \cong M \tag{3.67}$$

for sufficiently large i. A stable ratio would only be achieved for a strictly linear system. As Young has proved, the rate of approach to stationarity in equation (3.67) is much slower for systems with multiple large eigenvalues. Ralston (1965) describes a formula which accounts for the effect of two or more large and closely spaced eigenvalues which together dominate the convergence rate. Stationarity is achieved more slowly in the latter case than would be the case if only one dominant eigenvalue prevailed.

In nonlinear systems (equation (3.58)) the implied linearization from a succession of x vectors will change due to the nonlinear interactions among the iterate variables. Stationarity of successive error norms may only occur as one approaches the converged solution.

Assuming complete interaction among the iterate variables one can define the norm of the error vector in terms of the largest element of e rather than the Euclidian norm, and estimate M from

$$e_j^{(i+1)}/e_j^{(i)} = M \tag{3.68}$$

for monitoring purposes, where j is the index of the maximum element in the vector e. All errors will eventually change by the same ratio if the iteration is continued for long enough. Having estimated M following a few successive substitutions, and estimating subsequent iterations,

$$e^{(i+1)} = Me^{(i)}$$
$$x^{(i+2)} = x^{(i+1)} + Me^{(i)}$$
$$e^{(i+2)} = x^{(i+3)} - x^{(i+2)} = Me^{(i+1)} = M^2 e^{(i)}$$
$$x^{(i+3)} = x^{(i+1)} + Me^{(i)} + M^2 e^{(i)}$$

$$\cdot$$
$$\cdot$$
$$\cdot$$

$$x^* = x^{(i+1)} + e^{(i)}(M + M^2 + M^3 + \ldots) \tag{3.69}$$

Carrying out the estimated iteration procedure indefinitely, we find that the xs form a geometric progression. Subtracting $x^{(i)}$ from both sides of equation (3.69) gives

$$x^* - x^{(i)} = e^{(i)}(1 + M + M^2 + \ldots) \tag{3.70}$$

or, summing the progression,

$$x^* = x^{(i)} + \frac{x^{(i+1)} - x^{(i)}}{1 - M} \tag{3.71}$$

Equation (3.71) is the acceleration equation for DEM, x^* is the extrapolated value of the iterate vector based on the best estimate of M. From equation (3.58) we compute $x^{(i+1)}$ using $x^{(i)}$ as the estimate, and extrapolate to x^* using the same M on all iterate variables.

The acceleration factor for DEM

$$\omega_{\text{DEM}} = 1/(1-M) \tag{3.72}$$

is much larger than ω_o for JOE. It is equivalent to replacing m by M in equation (3.66). As pointed out by Orbach and Crowe, one cannot apply such a large ω at each subsequent iteration, otherwise the method becomes unstable (for example 1, $\omega_{\text{DEM}} = 10.0$, $M' = 11.0$). For DEM a number of direct substitution steps are taken before each new extrapolation is made.

One cannot use norms defined over the entire vector e for systems that are decoupled. In a completely decoupled system, where each iterate variable is independent, one would need to use a characteristic M for each variable. In flowsheeting problems the iterate variables are neither all strongly coupled nor decoupled. All species are weakly coupled, at least, through the phase equilibrium properties, and strongly coupled if they participate in chemical reactions with each other. In systems with more than one tear stream, there would be strong coupling between the same species variable in each tear stream.

Secant method

The secant (or Wegstein) method (SM) is in the same class as JOE and DEM. The secant method is inspired from the Newton–Raphson procedure but when applied to multidimensional systems in its one-dimensional form it is, in fact, a relaxation method. As we noted earlier, separate acceleration factors are applied to each variable. We have already proposed that this is essential for DEM and JOE if the iterate variables are uncoupled or weakly coupled.

The secant method is

$$x_j^* = x_j^{(i)} + \frac{x_j^{(i+1)} - x_j^{(i)}}{1-s_j} \tag{3.73}$$

for each iterate variable j, and

$$\omega_{\text{SM},j} = 1/(1-s_j) \tag{3.74}$$

The slope s_j is obtained from two successive trials with equation (3.58):

$$s_j = \frac{f_j(x^{(i+1)}) - f_j(x^{(i)})}{x_j^{(i+1)} - x_j^{(i)}} \tag{3.75}$$

So long as equation (3.75) is computed from successive substitution steps,

$$x_j^{(i+2)} = f_j(x^{(i+1)})$$

$$x_j^{(i+1)} = f_j(x^{(i)})$$

we have

$$s_j = \frac{x_j^{(i+2)} - x_j^{(i+1)}}{x_j^{(i+1)} - x_j^{(i)}} = \frac{e_j^{(i+1)}}{e_j^{(i)}} \tag{3.76}$$

Noting equations (3.67) and (3.72) along with the above, we see that SM is *entirely* equivalent to DEM, except for the choice to use the norm of e to define M rather than to define each s_j, as here.

SM in a multidimensional environment suffers from the same instability problems (in the light of relaxation theory) that one would experience with DEM if it were applied every iteration step, since acceleration factors much greater than ω_o are encountered in most problems. It has been the practice to bound SM in the following manner to control stability:

$$q_j = 1 - \omega_{\mathrm{SM},j} = s_j/(s_j - 1) \tag{3.77}$$

$$0 \geqslant q_j \geqslant -n$$

where the lower bound $-n$ is usually -2 to -5. The upper bound on q ensures that SM never interpolates ($q = 0$ is equivalent to $\omega = 1$), i.e. it defaults to successive substitution, and the lower bound limits extrapolation.

However, SM like DEM can benefit from delayed application to the point where the lower bound, at least, can be eliminated. Let us look again at example 2(a).

Example 2(b). SM with delay was applied to our pathological example 2(a). Four successive substitution (SS) steps were followed by one unbounded SM step, then two additional SS steps were taken. The results are shown in table 3.12. Obviously, on this linear system one SM step after four SS steps gets very close to the exact solution. The

Table 3.12. *Results for example 2(b)*

Step†	x_1	x_2
0	1.0	1.0
1 SS	83.99985	1.125×10^{-4}
2 SS	165.9957625	$3.150002813 \times 10^{-3}$
3 SS	247.9906124	$6.225077344 \times 10^{-3}$
4 SS	329.9844373	$9.300114846 \times 10^{-3}$
(s_j	0.9999874994	0.9999879582)
SM	79 996.16	83 044.06
4*SM	6 559 439	255.3715
Exact solution	6 561 476	246.0738
5 SS	6 559 439.077	245.997418
6 SS	6 559 439.078	245.997415

†SS, successive substitution; SM, secant method.

acceleration factors of 80 000 (versus JOE of 2.0) suggest that SM would be inherently unstable for this problem if the same ω_{SM} were to be applied every step. The nature of the problem precludes additional delayed SM steps since SS-step results are changing only in the tenth digit for x_1 and ninth digit for x_2. Most computers, using single-precision arithmetic, would be unable to handle the problem beyond the first delayed SM step.

Example 3. A problem by Cavett (1963) has been used repeatedly in the literature to study various acceleration methods, most recently by Rosen and Pauls (1977). The system is shown in figure 3.11, and Rosen and

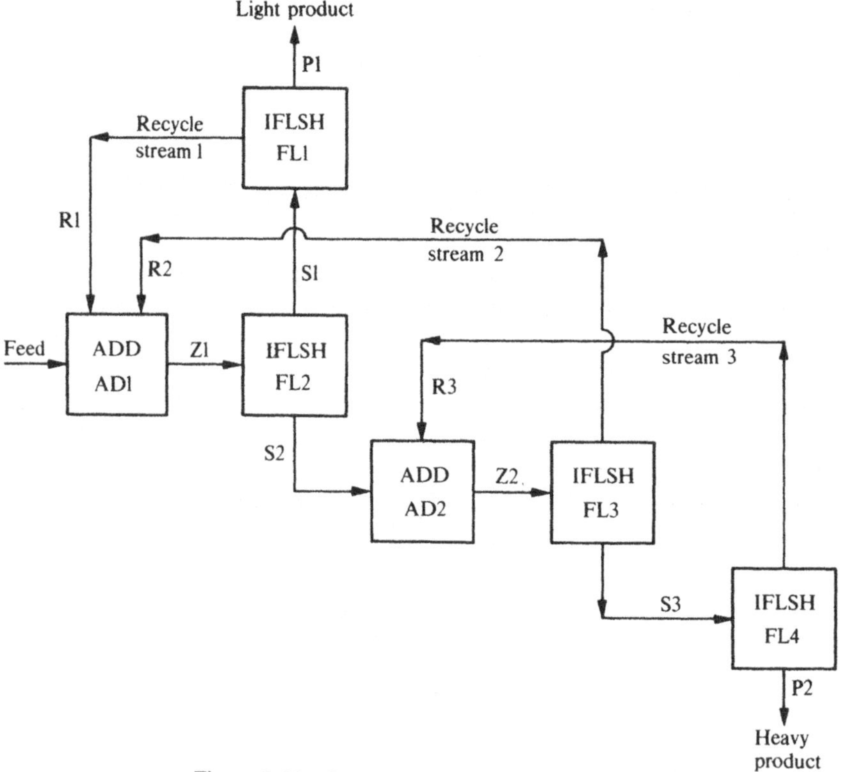

Figure 3.11. Cavett problem to study tearing.

Pauls' results in table 3.13. Figure 3.12 clearly indicates that the successive substitution approach takes a long time to stabilize. The Cavett problem was solved by Rosen and Pauls using two different sets of streams to be guessed and iterated, i.e. torn. One tear set was streams S1 and S3; the other was streams R1, R2 and R3. For either tear set the plot of log error versus iteration number does not become linear until

Table 3.13. *Data and solution for example of figure 3.11 (After Rosen and Pauls, 1977)*

Component	(gmoles/s)			K-values at solution			
	Feed	P1	P2	Flash 1	Flash 2	Flash 3	Flash 4
Nitrogen	45.23	45.23	0.000011	5.94	24.6	149.7	620.8
Carbon dioxide	626.97	626.86	0.60	1.51	4.64	21.1	72.3
Hydrogen sulphide	42.85	42.26	0.66	0.89	2.03	8.28	27.1
Methane	378.22	378.32	0.02	3.09	10.3	52.9	200.1
Ethane	302.46	364.04	1.83	1.00	2.66	11.2	39.3
Propane	289.27	239.79	51.04	0.502	0.943	3.29	10.8
Isobutane	76.28	25.03	51.81	0.310	0.445	1.34	4.22
n-Butane	194.43	37.80	157.50	0.246	0.342	0.99	3.07
Isopentane	99.80	4.70	95.19	0.155	0.164	0.417	1.22
n-Pentane	142.66	4.39	138.35	0.126	0.132	0.327	0.944
n-Hexane	222.82	1.33	221.54	0.064	0.051	0.107	0.290
n-Heptane	329.13	0.47	328.69	0.035	0.022	0.039	0.101
n-Octane	232.89	0.069	232.83	0.017	0.008	0.013	0.033
n-Nonane	210.73	0.015	210.72	0.009	0.004	0.005	0.012
n-Decane	105.01	0.002	105.01	0.005	0.002	0.002	0.004
n-Undecane	153.35	0.0008	153.35	0.003	0.0008	0.0009	0.002
Temperature (K)	322	311	303	311	322	309	303
Pressure (bar)	4.39	56.2	1.91	56.2	19.6	4.39	1.91

10–20 iterations, illustrating the effect of multiple large eigenvalues and/or nonlinearity. Also shown in figure 3.12 is the result of a bounded secant procedure

$$x_j^* = q_j x_j^{(i)} + (1-q_j)f_j(x_j^{(i)}) \tag{3.61}$$

with $0 \geqslant q_j \geqslant -5$ and acceleration every fourth iteration.

For our purposes we were not able to reproduce Rosen and Pauls' results on the computer, principally because we used a different program for estimating physical properties. However, by changing the flash equilibrium pressures to gauge pressures rather than absolute pressures, we were able to run the same feed mixture, using tear set (S1, Z2), and achieve convergence in 64 iterations to a relative error in all variables of 0.0005. This contrasts with the authors' results with (R1, R2, R3) which converged in about 54 iterations. Stream compositions are not identical, obviously, but the modified problem has the same characteristics. On our (perhaps tougher) problem it was found that the ratios of SS-step differences in the mass balance variables did not stabilize until the fifteenth iteration. However, a reasonable stability was observed after 8 iterations, although the maximum ratio between successive errors was only 70% of its ultimate value. A delayed SM was applied every eighth step, without lower bounding, and the system converged after 23 iterations with only two applications of the

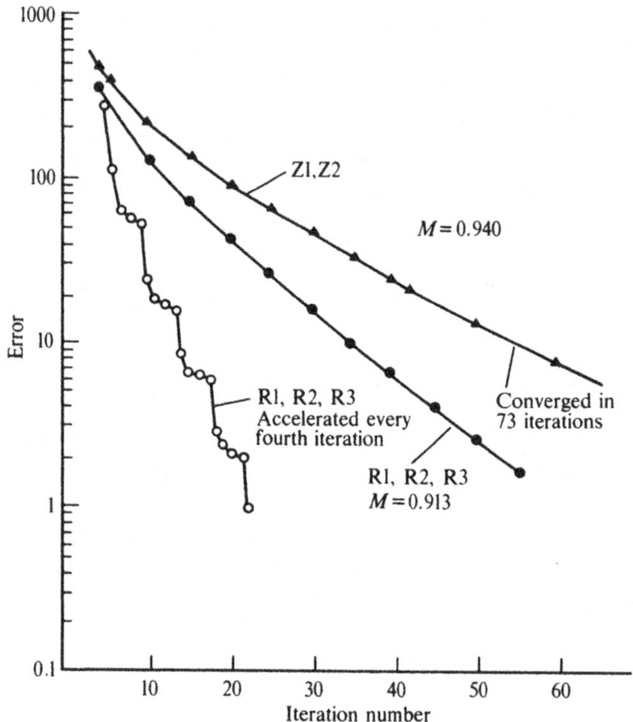

Figure 3.12. Convergence of the Cavett problem.

unbounded secant acceleration. The maximum ω_{SM} factor (i.e. $1-q$) in each application was substantially larger (as large as 25) than Rosen and Pauls' maximum of 6.

Other experiments on the same problem are summarized in table 3.14.

Observations

(1) JOE, DEM and SM are members of the same class of one-dimensional convergence-acceleration methods.

(2) SM is not analogous to the Newton–Raphson method (NRM) in coupled multidimensional systems although it approximates NRM in uncoupled systems.

(3) JOE places an approximate upper bound of 2 on the acceleration factors for one-dimensional methods which are applied every iteration. This is equivalent to q_j of -1 in the bounded SM.

(4) Experience indicates that the unbounded SM with delay is the best one-dimensional method, provided a sufficient delay is introduced

Table 3.14. *Convergence of Cavett's problem (relative error = 0.0005)*

| | q | | Number of iterations |
SM delay	Upper bound	Lower bound	to convergence
12	0	$-\infty$	27
8	0	$-\infty$	23
6	0	$-\infty$	22
5	0	$-\infty$	19
5	0	-10	27
5	0	-5	27
4	0	$-\infty$	47
4	0	-10	27
4	0	-5	27
SS			64

to stabilize the problem between SM steps. Very large acceleration factors may be used. Delays of three-to-five steps are typical with the larger delay for more difficult problems, i.e. problems with several large eigenvalues.

3.4 Solving sets of sparse nonlinear equations

In the previous section we considered methods for solving n nonlinear equations in $m(=n)$ unknowns. We did not explicitly take advantage of the sparse nature of these equations, a topic we now wish to cover. We first consider the particular case when $n = m$, i.e. the number of equations and unknowns are equal. We shall then consider the more general case when $m > n$, i.e. the number of variables exceeds the number of equations. For this case $m-n$ variables will have to be selected to be input variables, leaving us with n equations in n variables to solve. The selection of which variables should be input can profoundly affect the ease with which we can solve the equations, and, if we have no preference as to which are selected *a priori*, we can often take real advantage of this observation.

3.4.1 Solving square, sparse sets of equations

We shall define our equation set as being 'square' if we have n equations in exactly n unknowns. For a sparse, square set of equations one will often find that a subset of the equations can be found which involves k_1 equations in precisely k_1 unknowns. If such is the case, these k_1 equations can be solved first and by themselves for the k_1 variables, leaving us then to solve the remaining $n-k_1$ equations in $n-k_1$ unknowns. Again we may, among these remaining equations, discover

k_2 equations in exactly k_2 of the remaining variables, allowing us to solve these next and by themselves. Clearly this process may continue several times, decomposing our larger problem into a number of smaller ones which are solved in sequence. The decomposing of the problem is termed 'partitioning' (Harary, 1959) and the sequence to solve the partitions is termed a 'precedence order' for them.

An example will illustrate the ideas here. Suppose we wish to solve the following five equations, f_1 to f_5, in the five unknowns, x_1 to x_5.

$$f_1: \ x_1 + x_4 - 10 = 0$$
$$f_2: \ x_2^2 x_4 x_3 - x_5 - 6 = 0$$
$$f_3: \ x_1 x_2^{1.7}(x_4 - 5) - 8 = 0$$
$$f_4: \ x_4 - 3x_1 + 6 = 0$$
$$f_5: \ x_1 x_3 - x_5 + 6 = 0$$

Initially these equations look difficult to solve, but they have been contrived, and, in fact, are very simple to solve. We can expose the underlying structure for these equations by displaying their incidence (occurrence) matrix (see Himmelblau, 1973)

$$
\begin{array}{c}
\\ f_1 \\ f_2 \\ f_3 \\ f_4 \\ f_5
\end{array}
\begin{array}{c}
\begin{array}{ccccc} x_1 & x_2 & x_3 & x_4 & x_5 \end{array} \\
\left[
\begin{array}{ccccc}
1 & & & 1 & \\
& 1 & 1 & 1 & 1 \\
1 & 1 & & 1 & \\
1 & & & 1 & \\
1 & & 1 & & 1
\end{array}
\right]
\end{array}
$$

where each row corresponds to an equation and each column to a variable. A nonblank entry (here a '1') appears in row i and column j if variable x_j appears explicitly in equation f_i.

We can rearrange the rows and columns, that is, put them into a different order, to discover that this problem partitions and precedence orders rather nicely:

$$
\begin{array}{c}
\\ f_1 \\ f_4 \\ f_3 \\ f_5 \\ f_2
\end{array}
\begin{array}{c}
\begin{array}{ccccc} x_1 & x_4 & x_2 & x_3 & x_5 \end{array} \\
\left[
\begin{array}{ccccc}
1 & 1 & & & \\
1 & 1 & & & \\
1 & 1 & 1 & & \\
1 & & & 1 & 1 \\
& 1 & 1 & 1 & 1
\end{array}
\right]
\end{array}
$$

We note immediately that equations f_1 and f_4 involve only variables x_1 and x_4, and they can therefore be solved first:

$$f_1: \ x_1 + x_4 = 10$$
$$f_4: \ -3x_1 + x_4 = -6$$

giving $x_1 = 4$ and $x_4 = 6$. Since we know values for x_1 and x_4, equation f_3 can be solved by itself for x_2:

$$f_3: \quad x_1 x_2^{1.7} (x_4 - 5) - 8 = 4x_2^{1.7} (6-5) - 8 = 0$$

giving $x_2 = (8/4)^{1/1.7} = 1.5034$. Lastly the equations f_2 and f_5 can be solved for x_3 and x_5:

$$f_2: \quad x_2^2 x_4 x_3 - x_5 = (1.5034)^2 \cdot 6x_3 - x_5 = 6$$

$$f_5: \quad x_1 x_3 - x_5 = 4x_3 - x_5 = -6$$

giving $x_3 = 1.2550$ and $x_5 = 11.0202$.

Algorithms exist for partitioning a set of equations and then placing them into a precedence order. A principal requirement is that one must first assign, to each equation, a variable which occurs in it. Each variable must be assigned to exactly one equation. The assigned variable is termed an output variable for the equation, which implies that the equation could, in principle, be used to solve for that variable in terms of the others appearing in it. Steward (1962) proved that (1) the equations are 'structurally singular' if such an assignment proves impossible, i.e. one can find a subset of q equations involving p ($<q$) variables, making $q-p$ equations either redundant and/or inconsistent with the remaining p equations; and that (2) if the equations fully partition, i.e. each partition involves but one equation and variable, then the output assignment is unique; conversely, if the output assignment is not unique, a partition with two or more equations and variables in it must exist.

The output assignment problem is perhaps most efficiently handled as an 'assignment problem' in linear programming (Gupta *et al.*, 1974). For our example set of equations the following represents such an assignment:

$$
\begin{array}{c}
\begin{array}{ccccc}
x_1 & x_2 & x_3 & x_4 & x_5
\end{array} \\
\begin{array}{c} f_1 \\ f_2 \\ f_3 \\ f_4 \\ f_5 \end{array}
\left[
\begin{array}{ccccc}
① &   &   & 1 &   \\
  & 1 & 1 & 1 & ① \\
1 & ① &   & 1 &   \\
1 &   &   & ① &   \\
1 &   & ① &   & 1
\end{array}
\right]
\end{array}
$$

For a small set of equations one can usually find such an assignment by trial and error. We shall examine a more difficult example shortly.

The equations and variables may now be represented as a directed graph as follows. For each equation f_i draw a node. Draw a directed edge from the node f_i to f_j if, and only if, the assigned output variable for f_i appears in f_j. Thus the node for f_1 above will have directed edges to the nodes for f_3, f_4 and f_5. The node for f_2 (note that x_5 is the assigned output variable) will have a directed edge from it to f_5, and so forth for the other equations. (Figure 3.13 illustrates our completed graph.) The graph so produced represents an information flow diagram. If f_1 is used to calculate x_1 (its assigned output variable), then the value for this variable is required by each of the equations f_3, f_4 and f_5 before they can be used to calculate their assigned output variables. We can now partition and precedence order this graph fairly readily.

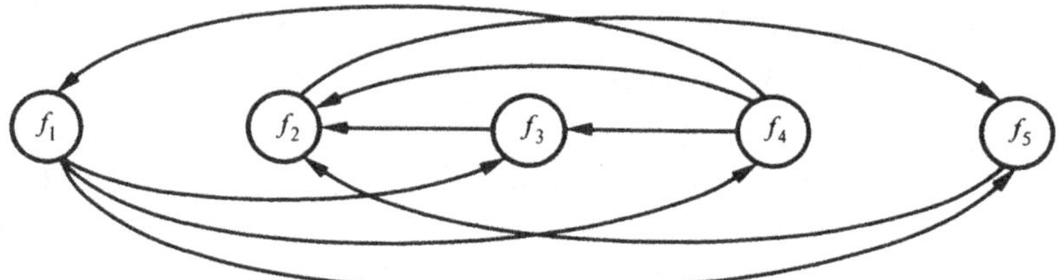

Figure 3.13. Directed graph for five-equations example.

The most efficient algorithm for this purpose is by Tarjan (1972) (see Duff and Reid, 1976). For our purposes the following algorithm (Sargent and Westerberg, 1964) is only slightly less effective for most problems, is quite similar in nature, but is much easier to present. One can partition and precedence order these equations by tracing equation outputs. One traces from an arbitrary first equation, say f_1, forming a string of nodes until (*a*) a node appears which has no output, or (*b*) a node appears which is already on our string. For the first case (case (*a*)) we add this node with no output edge to the top of a list (let's call it 'list 1') and delete the node and all edges into and out of it from the graph. For case (*b*) we merge all nodes appearing between the occurrence of the repeated node with the repeated node into a new 'group' node. This group node is treated in exactly the same manner as the original nodes with the process of tracing nodes then continuing from it. (The Tarjan algorithm cleverly avoids this grouping until the end.)

For our example we start the string with f_1 and proceed as follows:

(1)   String: $f_1 \rightarrow f_3 \rightarrow f_2 \rightarrow f_5 \rightarrow f_2$ case (*b*)

Merge f_2 and f_5 to form a group node.

(2)   String: $f_1 \rightarrow f_3 \rightarrow (f_2 f_5)$

We now attempt to extend the string and discover that the group node $(f_2 f_5)$ has no output edges.

(3)   String: $f_1 \rightarrow f_3 \rightarrow (f_2 f_5)$ case (*a*)

We place $(f_2\ f_5)$ onto the top of list 1, deleting f_2 and f_5 and all edges into and out of them from our problem. Again, our attempt to extend the string fails as f_3 has no output edges remaining.

(4)   String: $f_1 \rightarrow f_3$ case (a)

We place f_3 onto the top of list 1, deleting it and all its edges from the graph. We again trace outputs from f_1 getting

(5)   String: $f_1 \rightarrow f_4 \rightarrow f_1$ case (b)

Merging f_1 with f_4 and continuing gives

(6)   String: $(f_1 f_4)$ case (a)

We place $(f_1 f_4)$ onto the top of list 1 and delete the node $(f_1 f_4)$ from our system. No nodes remain and our string is empty so we terminate, having created the following list 1:

$$
\begin{array}{ll}
(f_1 f_4) & \text{partition 1} \\
f_3 & \text{partition 2} \\
(f_2 f_5) & \text{partition 3}
\end{array}
$$

We find we have found exactly the partitioning and precedence order discovered earlier. Figure 3.14 traces the above steps as they occur on the directed graph.

Steward (1962) proved that partitioning is unique, by showing that the only possible alternative output variable assignments must arise from, and only from, shuffling the assignments among the variables and equations within a partition. Precedence ordering, on the contrary, need not be unique, as two partitions may exist which may be solvable in either order or, indeed, in parallel. Figure 3.15 illustrates such an example. The third partition does not require values for the variables x_2 and x_3 which appear in the second partition.

We now return to the problem of assigning output variables to equations. The example in figure 3.16(a) is an interesting one in that it is a small problem, but it will defeat virtually all 'quick and dirty' schemes normally invented to get an assignment. The usual approach is to select first the column (or row) with fewest incidences. The row assigned to that column is among those in the column and itself has the fewest incidences in it. Following this general scheme we shall assign column 1 first since it has only two incidences in it. We select row 8 because it too has only two incidences in it, rather than row 7 which has three. We delete the assigned row and column and repeat until, unfortunately, we are left with the assignment shown in figure 3.16(b). Only equation f_7 and variable x_9 remain, and variable x_9 does not appear in equation f_7.

At this point we must try to find a so-called 'Steward path' leading between any incidence in the row for equation f_7 to any incidence in the column for variable x_9. Figure 3.16(b) illustrates such a path. It starts with an unassigned incidence, moves to an assigned one, changes direction by 90°, locates an unassigned incidence, changes direction

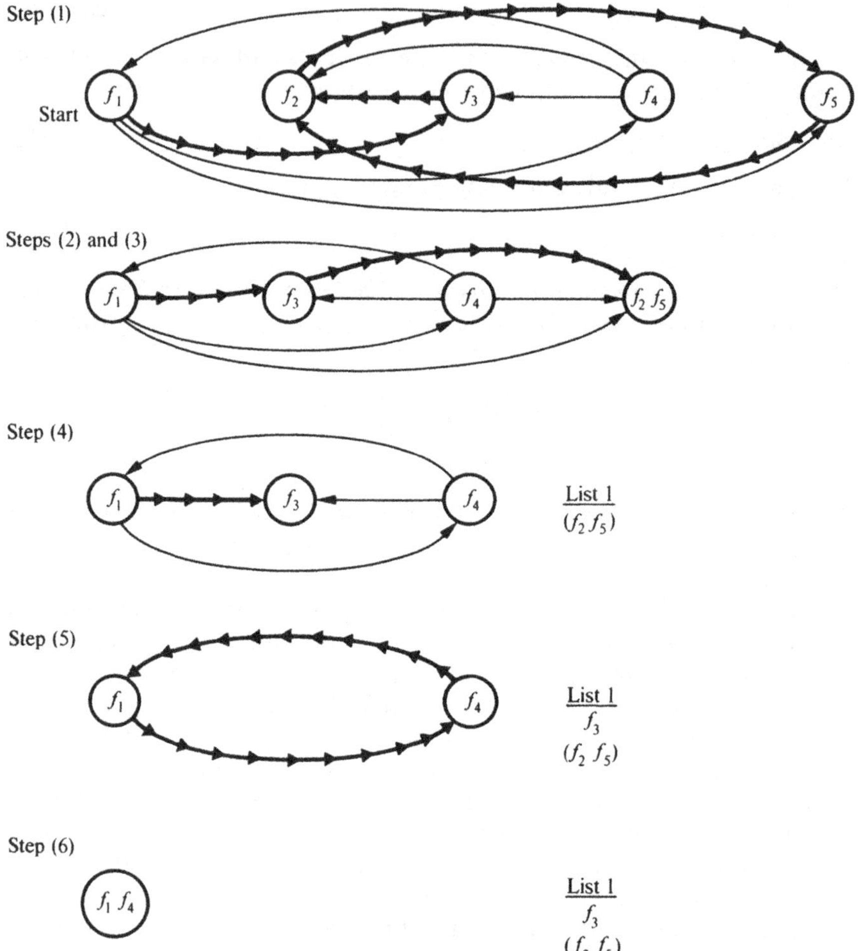

Figure 3.14. Partitioning and precedence ordering the five-equations example. The bold lines represent the string tracing occurring for the step indicated.

again by 90°, moves to an assigned incidence, etc. Assigned and unassigned incidences alternate along such a path, with the first and last being unassigned. The path has one more unassigned incidence along it than assigned. Also each row and column included in the path has only one assigned and one unassigned incidence in it. We can therefore interchange the assignments, unassigning the currently assigned incidences and assigning the currently unassigned. We will have created one additional assignment in this manner. A detailed algorithm for finding Steward paths is given by Gupta (1972).

Figure 3.16(*c*) has the reassignments made, and we have a fully assigned set of equations which can now be partitioned and precedence ordered. Finding Steward paths for moderately large problems, say 25 equations, is not too difficult to do by trial and error by hand. If no such path exists when one is needed, no output assignment exists and the equations are structurally singular, as mentioned earlier.

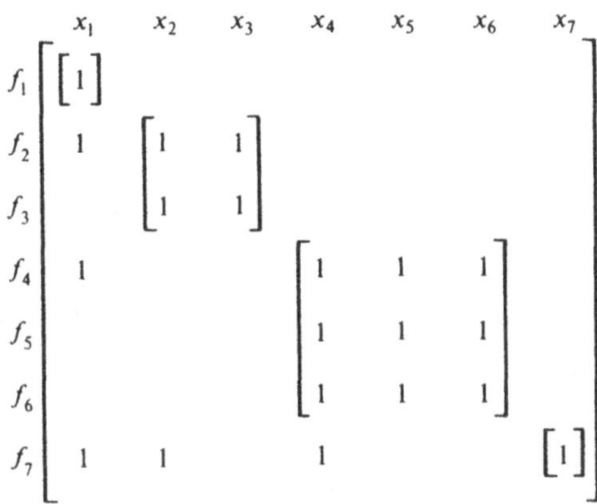

Figure 3.15. Incidence matrix for example with a nonunique precedence order.

The solving of sparse, square equation sets may of course be done by direct linearization. The sparseness of the equations will effectively be accounted for by solving the resulting sparse linear equations using techniques such as those outlined in section 3.2.3. One can, of course, partition and precedence order the equations first, as indicated in the previous section. Each partition can be then solved in sequence using direct linearization techniques as discussed in section 3.3.1. Three advantages to partitioning are apparent. First, of course, a large problem may become a series of small problems. Secondly, the initial values need only be guessed for the first partition before it is solved. Then actual values exist for these variables which may permit a better guess for values for the second partition and/or aid in developing a numerically superior pivot sequence for the second partition. Thirdly, if the problem is partitioned and a pivot sequence is then found only within the partitions, the fill-in obtained (which results in more arithmetic operations) will only occur within the partitions. If not partitioned, one may select pivots not within the partitions and will very likely create more nonzero elements this way.

One can also solve sparse, square equation sets using tearing methods. The ideas behind tearing are perhaps best introduced by example, and in fact, we have already seen examples of it in this book.

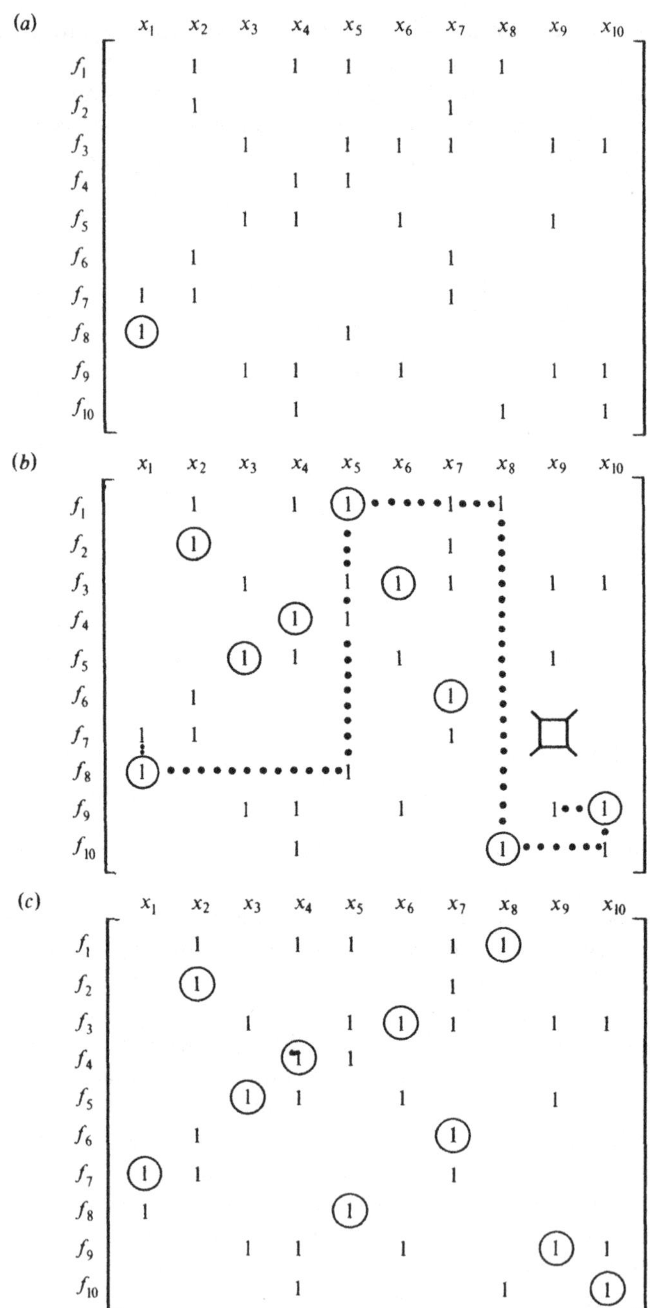

Figure 3.16. Example problem with difficulty encountered when assigning output variables. (*a*) First assignment. (*b*) Steward path from column 9 to row 7 illustrated – note that assignment of x_9 to f_7 is not possible. (*c*) Reassigned and complete output assignment.

In section 3.1.1 we set up and solved the equations for a simplified three-component isothermal flash unit. The number of equations solved was 14; these included three material balance equations, three equilibrium relationships, six physical property relationships, plus two others. Reviewing the algorithm, we had in it a loop of calculations where we iterated the vapour fraction to drive the expression $\Sigma x_i - \Sigma y_i$ to zero; within this loop were the instructions:

3.   Guess V/F; then $V = (V/F) \cdot F$

4. $(a)\, L = F - V$
     $(b)\, x_i = z_i F/(K_i V + L)\quad i = 1, 2, 3$
     $(c)\, y_i = K_i x_i\quad i = 1, 2, 3$

5.   Evaluate $\Sigma x_i - \Sigma y_i$
     (a) If not essentially zero, iterate
     (b) Otherwise, continue

There are eight equations in eight unknowns in this loop, and we are solving them by iterating only one variable. This approach is the essence of tearing, where the variables to be iterated, V/F in this case, are called 'tear' variables and are many fewer in number than the total number of variables. The equations within the iteration loop are said to be a 'group' of equations since they must be solved simultaneously.

Before continuing, let us also investigate a second example of tearing that we have already presented. To solve a flowsheet using the sequential modular approach, we illustrated an example where a recycle stream was guessed. The flowsheet units could then be solved in sequential order until the unit producing the recycle was encountered. At this point, we compared the guessed and calculated values for the recycle stream and iterated if they did not agree. The recycle variables are tear variables here, and the calculations for the units within the loop are a group of equations to be solved simultaneously.

It should be obvious that the particular tear variables used vary with the approach devised. Thus we can 'play' with the equations and attempt to derive more appealing approaches for solving. Some interesting questions arise. Can the process of finding tear variables be automated? If so, what limitations exist? Will the resulting calculations converge? We can discover some ideas relating to these questions by reinvestigating the equations in section 3.1.1 for the simple isothermal flash unit. First, they have considerable sparseness, which can be illustrated by the relevant incidence matrix, figure 3.17. We defined an incidence matrix earlier, but, to repeat, the rows represent the equations and the columns the variables for our problem. A blank entry in row i and column j denotes that equation i does not explicitly contain variable j; a nonblank entry denotes that it does. Function f_1 corresponds then to

	y_1	y_2	y_3	x_1	x_2	x_3	z_1	z_2	z_3	V	L	F	K_1	K_2	K_3	P_1^0	P_2^0	P_3^0	P	T
(1) f_1	1			1			1			1	1	1								
(1) f_2		1			1			1		1	1	1								
(1) f_3			1			1			1	1	1	1								
(2) f_4										1	1	1								
(3) f_5	1			1									1							
(3) f_6		1			1									1						
(3) f_7			1			1									1					
(4) f_8													1			1			1	
(4) f_9														1			1		1	
(4) f_{10}															1			1	1	
(5) f_{11}																1				1
(5) f_{12}																	1			1
(5) f_{13}																		1		1
(6) f_{14}	1	1	1	1	1	1														
(7) f_{15}							1	1	1											

Figure 3.17. Incidence matrix for a simple 20 variable, 15 equation flash calculation (the relevant equation number for each function is given in parentheses).

the particular equation

$$y_1 V + x_1 L = z_1 F$$

and nonblank entries appear in the row for f_1 in six columns – under y_1, V, x_1, L, z_1 and F.

We are now able to see the 'structure' of the equation set with which we are dealing. We have listed 15 functions in 20 variables. Thus five variables may have values chosen for them. We wish to solve an isothermal flash calculation with the five variables specified being F, z_1, z_2, P and T; thus these variables may be treated as fixed and deleted from our problem. Figure 3.18 gives the remaining incidence matrix, which has 15 equations in 15 unknowns.

We can now partition and precedence order the equations to obtain the rearranged matrix shown in figure 3.19. We note that the variables P_1^0, P_2^0, P_3^0, z_3, K_1, K_2 and K_3 can be calculated in that order and one at a time. We are left with a final partition involving eight variables and eight equations. Figure 3.20 shows this partition with all other equations and variables deleted.

It is now possible to conclude that no solution procedure exists which will not require us to solve these last eight equations simultaneously, or require us to do some algebraic rearrangement. We might hope that they are totally linear, but it is obvious that they are not, as some equations involve product terms in the unknowns remaining.

We now wish to develop a solution procedure for these eight equations in eight unknowns. A fairly effective and simple approach is to attempt to create, by 'tearing', an equation with only a single variable in it. Suppose we are willing to guess the values for one or more of the

	y_1	y_2	y_3	x_1	x_2	x_3	z_3	V	L	K_1	K_2	K_3	P_1^0	P_2^0	P_3^0
f_1	1			1				1	1						
f_2		1			1			1	1						
f_3			1			1		1	1						
f_4								1	1						
f_5	1			1						1					
f_6		1			1						1				
f_7			1			1						1			
f_8										1			1		
f_9											1			1	
f_{10}												1			1
f_{11}													1		
f_{12}														1	
f_{13}															1
f_{14}	1	1	1	1	1	1									
f_{15}							1								

Figure 3.18. Incidence matrix after specified variables have been deleted.

	P_1^0	P_2^0	P_3^0	z_3	K_1	K_2	K_3	y_1	y_2	y_3	x_1	x_2	x_3	V	L
f_{11}	1														
f_{12}		1													
f_{13}			1												
f_{15}				1											
f_8	1				1										
f_9		1				1									
f_{10}			1				1								
f_1								1			1			1	1
f_2									1			1		1	1
f_3										1			1	1	1
f_4														1	1
f_5					1			1			1				
f_6						1			1			1			
f_7							1			1			1		
f_{14}								1	1	1	1	1	1		

Figure 3.19. Incidence matrix after partitioning and precedence ordering of equations.

	y_1	y_2	y_3	x_1	x_2	x_3	V	L
f_1	1			1			1	1
f_2		1			1		1	1
f_3			1			1	1	1
f_4							1	1
f_5	1			1				
f_6		1			1			
f_7			1			1		
f_{14}	1	1	1	1	1	1		

Figure 3.20. Incidence matrix after assigning and deleting functions $f_{11}, f_{12}, f_{13}, f_{15}, f_8, f_9$ and f_{10}, and variables $z_3, K_1, K_2, K_3, P_1^0, P_2^0$ and P_3^0.

variables. Then we could consider these 'guessed variables' as known, deleting them from the problem. Of course we wish to guess the fewest number possible. The variables we select to guess will, of course, have to be iterated, as it is unlikely that we can initially guess their values correctly. These variables are called tear variables. We proceed.

Function f_4 has only two variables (as do f_5, f_6 and f_7). We can decide to use it to calculate L and require at this point that we need to guess a value for V. Thus we identify V as a tear variable. Now the last two columns can be deleted along with the fourth row, leaving the matrix in figure 3.21(a). We observe immediately a rather interesting thing. All

(a)

	y_1	y_2	y_3	x_1	x_2	x_3
f_1	1			1		
f_2		1			1	
f_3			1			1
f_5	1			1		
f_6		1			1	
f_7			1			1
f_{14}	1	1	1	1	1	1

(b)

	y_1	x_1	y_2	x_2	y_3	x_3
f_1	1	1				
f_5	1	1				
f_2			1	1		
f_6			1	1		
f_3					1	1
f_7					1	1
f_{14}	1	1	1	1	1	1

Figure 3.21. Final incidence matrix. All functions are linear. Matrix (b) is matrix (a) rearranged to display structure.

the remaining equations are linear in all their *unknown* variables. Thus we could simply hand them over to a sparse matrix routine for solving. Note that the set is not square; there are seven equations in six unknowns so the sparse matrix routine will need to identify a 'redundant' equation and simply evaluate it. Its value is zero if V has been given a correct value; otherwise it is the 'error' which we drive to zero by adjusting V.

Rather than handing the equations in total to a sparse matrix routine, we can rearrange the order for the variables and equations to give the pattern in the incidence matrix in figure 3.21(b). Here it is obvious we have a number of linear equation pairs to solve for x_1, y_1, for x_2, y_2 and for x_3, y_3. Then f_{14} is simply evaluated to see if it is zero. We can show that solving these equation pairs corresponds precisely with the algebra we did to write equation (3.15).

Functions f_1 and f_5, for example, are

$$f_1: \ y_1 V + x_1 L = z_1 F$$

$$f_5: \ y_1 = K_1 x_1$$

At this point in the computation V, L, z_1, F and K_1 have values, so we are left with two linear equations in the unknowns x_1 and y_1:

$$Vy_1 + Lx_1 = z_1 F$$
$$y_1 - K_1 x_1 = 0$$

(3.78)

A simple Gaussian elimination to eliminate the y_1 term in the second equation gives

$$-(VK_1 + L)x_1 = -z_1 F$$

which is equivalent to equation (3.15). Thus we have been led to derive the intuitive solution discovered earlier. The approach we have taken can be formalized and made into an algorithm which may then be programmed on the computer. The input to the routine would be a set of equations, and the output a solution procedure for them: the order in which to solve each equation, which variable to solve for in each, the equation groups and, within them, the tear variables, and so forth.

The routine just alluded to is in practice enormously complex. It has to be designed to handle several thousand equations; this means that every feature of the incidence matrix, which we casually observed for our simple example, potentially becomes a huge combinatorial problem to locate. If solved without due care, it could lead to impossible computer time requirements. Thus clever means are crucial to store data, to search them, and so forth. The research literature abounds with these ideas and comparisons thereof, justifiably so if the problem is to be solved realistically.

Some of the references which give algorithms useful for solving square, sparse equation sets by tearing include the following: Steward (1965), Ledet and Himmelblau (1970), Westerberg and Edie (1971a, b), Barkley and Motard (1972), Kevorkian and Snoek (1973), Pho and Lapidus (1973), Soylemez and Seider (1973), Cheung and Kuh (1974), Guardabassi (1974), Stadtherr, Gifford and Scriven (1974), Kevorkian (1975), and Smith and Walford (1975).

3.4.2 Solving nonsquare, sparse sets of equations by tearing methods

Not all equation sets are 'square' when the user has finished supplying them. Not being square means the number of equations, n, is not equal to the number of variables, m. If more variables exist than equations, $m - n$ degrees of freedom exist for the problem, and $m - n$ additional specifications must be provided. The routines for deriving a solution procedure can handle this problem by strategically choosing $m - n$ variables to be decision variables whose values must then be supplied before the procedure is executed. If we examine the flash unit equations (3.7) to (3.13) of section 3.1.1 (assuming three components as we did in this section), we find five additional specifications are needed. We were solving an isothermal flash with a given feed so the specifications were for z_1, z_2, P, T and F as listed in 'equations' (3.14).

Suppose the flash unit is to be optimized to meet some economic or operational objective, and, in fact, the only variables we wish to fix *a priori* are z_1, z_2 and P. Then the solution procedure will have to find two variables for us to adjust when optimizing; these will be the decision variables. But which two? If we do not have any preferences, then why not find two which will ease the problem of solving the equations? For this problem the isothermal flash calculations prove to be as easy to solve as any others, so the two variables which would be chosen would very likely be F and T. Often, however, unexpected choices result, and the computational procedure derived for the best choice may contain no iteration loops, obviously a very desirable result.

The capability of finding decision variables is very useful and algorithms exist in the literature to solve this problem. Finding decision variables often significantly complicates the problem of finding the solution procedure if one looks extensively to find a best set of decision variables. The problem tends to be combinatorial.

We can illustrate a simple and relatively effective algorithm for this type of problem. The literature contains related algorithms for this problem (see, e.g., Christensen, 1970; Edie and Westerberg, 1971; Ramirez and Vestal, 1972). The example problem to be considered is the two-stage, three-component absorber shown in figure 3.22. We shall assume the pressure P and the liquid flowrate L are specified. The equations we shall use to model this unit are (for $i = 1, 2, 3$ and $j = 1, 2$).

Material balances:

$$x_{i,j-1}L + y_{i,j+1}V = x_{i,j}L - y_{i,j}V \qquad (3.79)$$

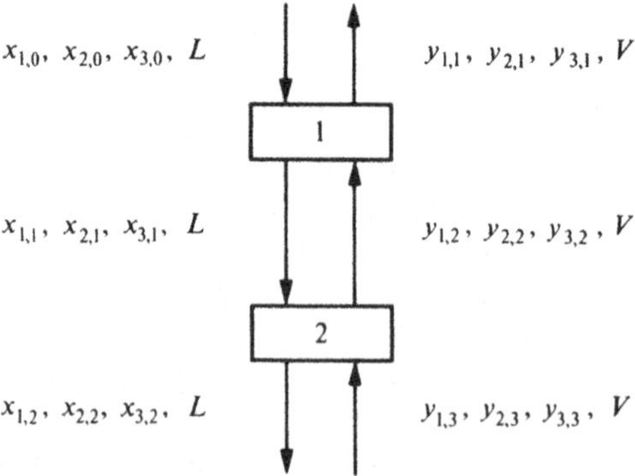

Figure 3.22. Two-stage, three-component absorber example.

Equilibrium:

$$y_{i,j} = K_{i,j} x_{i,j} \tag{3.80}$$

Physical property relationships (we assume no composition dependence):

$$K_{i,j} = K_{i,j}(T,P) \tag{3.81}$$

Other:

$$\sum_{k=1}^{3} x_{k,l} = 1 \qquad l = 0, 1, 2 \tag{3.82}$$

$$\sum_{k=1}^{3} y_{k,m} = 1 \qquad m = 1, 2, 3 \tag{3.83}$$

Figure 3.23 is an incidence matrix for these 24 equations in 28 variables. The two preselected decision variables, L and P, reduce our system to 24 equations in 26 variables; two more decision variables are needed.

We cannot partition as we did before when the set of equations was square. We must therefore proceed in a different manner. As we argued

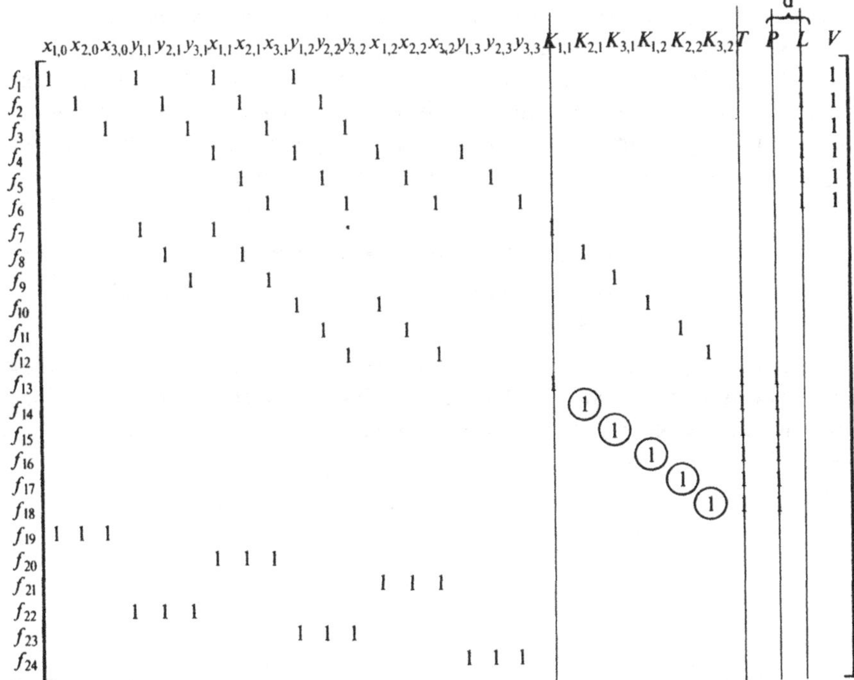

Figure 3.23. Incidence matrix for absorber example. d, decision variables.

earlier in section 3.4.1 when deriving a solution procedure for the flash unit, we need to find an equation with only one variable in it to get started. We can also look for a variable appearing in only one equation. For this latter case, that variable can be assigned (if we like the assignment) to that equation and may be identified as the last equation to be solved. Its output variable cannot be required as an input elsewhere so that variable can clearly be the last variable calculated.

For our example, no equations have only one variable, and no variables appear in only one equation. We must perforce 'create' an equation with only one variable in it (or we must 'create' a variable which appears in only one equation) if we are to proceed.

We would, of course, like to create as few such situations as possible because each one in excess of two (two of the set created will be selected to be decision variables) will lead to our having to guess a value for the variable and then iterate on this guess, i.e. we will create tear variables.

To minimize the number of potential tear and decision variables created, we search for an equation with the fewest variables. All but one of the variables will have to be known to permit the remaining one to be calculated. Looking at the equations, we see that several, f_{13} to f_{18}, contain only two variables. In fact, we see that if a value for T were known, either by guessing it or selecting it to be a decision variable, we could calculate all the $K_{i,j}$ variables. We could therefore pick T to be a 'known' variable and assign the $K_{i,j}$ variables as outputs from equations f_{13} to f_{18}. However we shall, for consistency with the rest of this problem, avoid making quite this commitment. Instead, we shall select one of the equations, say f_{13}, and simply list, as follows, both variables appearing in it. Our list, which we shall be creating as we proceed, starts as follows:

$$(1) \quad L, P \qquad \textit{a priori} \text{ decisions}$$
$$(2) \quad T, K_{1,1}$$
$$(3) \quad f_{13}$$

If a variable appears on our list it means that its value is known, either as a decision or a tear variable, or calculated. With T 'known' we see that we can assign $K_{2,1}$ to $K_{3,2}$ as output variables for equations f_{14} to f_{18}. We add these to our list:

$$(4) \quad (f_{14}, K_{2,1})$$
$$(5) \quad (f_{15}, K_{3,1})$$
$$(6) \quad (f_{16}, K_{1,2})$$
$$(7) \quad (f_{17}, K_{2,2})$$
$$(8) \quad (f_{18}, K_{3,2})$$

The above list is being formed with the following ideas in mind. First we shall list a variable by itself when a need for its value is required to permit further calculations. In the above we needed a value for T to proceed. Actually, in principle, we needed a value for T *or* $K_{1,1}$, since the other should be immediately available from equation f_{13}. We know even now in the analysis that we prefer to calculate $K_{1,1}$ using f_{13} rather than calculating T from that equation, but we shall not yet make this commitment. So we list both, and follow this listing with f_{13}.

We shall always list an equation as soon as all of its variables have appeared on our list, whether we have just created that situation or it has happened by itself in the course of the analysis. The equation, when listed, can be used at that point, perhaps as an error function into which we substitute all the current variable values and then check to see if the equation is, or is not, satisfied. We shall also list an equation and its assigned output variable if the assignment is forced because the equation is left with only one variable in it which is not yet listed. Items (4) to (8) appear on our list for just this reason.

As each variable and equation is listed, we can delete it from the incidence matrix. Figure 3.24 shows our current reduced incidence matrix. We proceed from this point.

	$x_{1,0}$	$x_{2,0}$	$x_{3,0}$	$y_{1,1}$	$y_{2,1}$	$y_{3,1}$	$x_{1,1}$	$x_{2,1}$	$x_{3,1}$	$y_{1,2}$	$y_{2,2}$	$y_{3,2}$	$x_{1,2}$	$x_{2,2}$	$x_{3,2}$	$y_{1,3}$	$y_{2,3}$	$y_{3,3}$	V
f_1	1									1									1
f_2		1									1								1
f_3			1			1			1			1							1
f_4										1			1			1			1
f_5											1			1			1		1
f_6									1			1			1			1	1
~~f_7~~																			
~~f_8~~																			
f_9						1			1										
f_{10}										1			1						
f_{11}											1			1					
f_{12}												1			1				
f_{19}	1	1	1																
f_{20}									①										
f_{21}													1	1	1				
f_{22}						①													
f_{23}										1	1	1							
f_{24}																1	1	1	

Figure 3.24. Incidence matrix for absorber example after first reduction.

We note now that if either $x_{1,1}$ or $y_{1,1}$ were known, the other could be calculated using equation f_7. Similarly if either $x_{2,1}$ or $y_{2,1}$ were known, the other could be calculated using equation f_8. We shall not make a commitment at this point as to which we shall assume is known and which to be calculated; rather we shall simply add them to our list:

$$(9) \quad x_{1,1}, y_{1,1}$$
$$(10) \quad f_7$$
$$(11) \quad x_{2,1}, y_{2,1}$$
$$(12) \quad f_8$$

Deleting these four variables and two equations allows us to discover that equation f_{22} may be solved for $y_{3,1}$ and equation f_{20} for $x_{3,1}$. We add these to our list

$$(13) \quad (f_{22}, y_{3,1})$$
$$(14) \quad (f_{20}, x_{3,1})$$

and delete them from the incidence matrix; figure 3.25 shows our current reduced incidence matrix.

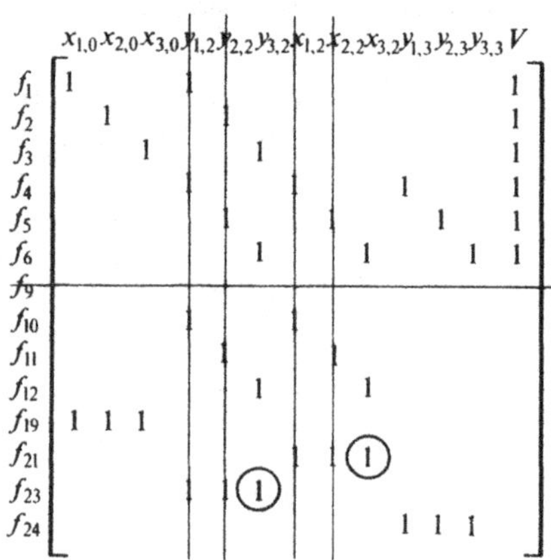

Figure 3.25. Incidence matrix after second reduction.

We note that f_9 has no associated variables; we delete it and add it to our list.

$$(15) \quad f_9$$

We note that either $y_{1,2}$ or $x_{1,2}$, if known, will permit the other to be calculated using f_{10}; $y_{2,2}$ and $x_{2,2}$ in f_{11} can be similarly treated.

$$(16) \quad y_{1,2}, x_{1,2}$$
$$(17) \quad f_{10}$$
$$(18) \quad y_{2,2}, x_{2,2}$$
$$(19) \quad f_{11}$$

We find now that f_{21} contains only $x_{3,2}$ and f_{23}, variable $y_{3,2}$. We assign them, delete them, and add them to our list.

$$(20) \quad (f_{21}, x_{3,2})$$
$$(21) \quad (f_{23}, y_{2,3})$$

Figure 3.26 shows our latest reduced matrix. We note f_{12} has no associated variables, and it is deleted and listed

(22) f_{12}

Finding f_1 with only two variables in it, we list the variables as known, and then f_1:

(23) $x_{1,0}$, V
(24) f_1

At this point equations f_2 and f_3 can be assigned,

(25) $(f_2, x_{2,0})$
(26) $(f_3, x_{3,0})$

leaving f_{19} with no variables:

(27) f_{19}

	$x_{1,0}$	$x_{2,0}$	$x_{3,0}$	$y_{1,3}$	$y_{2,3}$	$y_{3,3}$	V
f_1	1						1
f_2		1					1
f_3			1				1
f_4				1			1
f_5					1		1
f_6						1	1
~~f_{12}~~							
f_{19}	1	1	1				
f_{24}				1	1	1	

Figure 3.26. Incidence matrix after third reduction.

Finally, f_4 to f_6 can be assigned output variables $y_{1,3}$ to $y_{3,3}$, leaving f_{24} with no unlisted variables:

(28) $(f_4, y_{1,3})$
(29) $(f_5, y_{2,3})$
(30) $(f_6, y_{3,3})$
(31) f_{24}

Figure 3.27 gives our final list.

We now need to analyse this list to see what it is telling us. For example, an unassigned variable appearing on the list must be either a decision variable or it must be calculated by an unassigned equation which follows it. The reason why the equation must follow is that we have always listed an unassigned variable before any equations in which it appears. We also listed it only when its value was needed.

We carefully handled the placing of unassigned and assigned equations on the list, placing them there as soon as they could be used.

We must now assign an unassigned variable to each unassigned equation; we must also prescribe a role to each unassigned variable. This role may be a simply assigned variable, a tear variable or a decision variable. We can select the role to simplify the calculations required to evaluate all but the two variables which become decision variables.

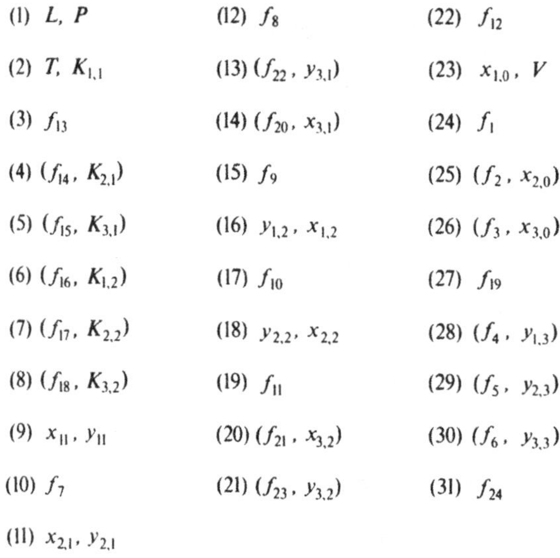

(1) L, P	(12) f_8	(22) f_{12}
(2) $T, K_{1,1}$	(13) $(f_{22}, y_{3,1})$	(23) $x_{1,0}, V$
(3) f_{13}	(14) $(f_{20}, x_{3,1})$	(24) f_1
(4) $(f_{14}, K_{2,1})$	(15) f_9	(25) $(f_2, x_{2,0})$
(5) $(f_{15}, K_{3,1})$	(16) $y_{1,2}, x_{1,2}$	(26) $(f_3, x_{3,0})$
(6) $(f_{16}, K_{1,2})$	(17) f_{10}	(27) f_{19}
(7) $(f_{17}, K_{2,2})$	(18) $y_{2,2}, x_{2,2}$	(28) $(f_4, y_{1,3})$
(8) $(f_{18}, K_{3,2})$	(19) f_{11}	(29) $(f_5, y_{2,3})$
(9) x_{11}, y_{11}	(20) $(f_{21}, x_{3,2})$	(30) $(f_6, y_{3,3})$
(10) f_7	(21) $(f_{23}, y_{3,2})$	(31) f_{24}
(11) $x_{2,1}, y_{2,1}$		

Figure 3.27. List of variables and equations discovered when analysing the incidence matrix. The equation order here is the final equation order to be used when solving the equations.

We will select two of the unassigned variables to be decision variables, but which two? We note the following observations to aid us. First of all, several of the unassigned variables are of our own creation as we could have assigned them earlier. For example, we find T and $K_{1,1}$ unassigned as item (2) but immediately following is f_{13}, an equation ideally suited to calculate $K_{1,1}$ if the value of T is known. A second observation is that a tear variable must precede its tear function because we must first guess the tear variable and then eventually evaluate an error function, i.e. an associated tear function. Also the tear variable value must affect the value of the tear function, so we must establish a cause-and-effect relationship between the two. We might also desire that the tear function explicitly contain the tear variable, for then we can create an explicit iteration loop.

Just how can we guarantee the cause-and-effect relationship? Structurally, it is not too difficult. We can take our original incidence matrix, and, for each assigned variable, perform a logical Gaussian elimination, noting only those locations which would, in general, become nonzero if we were solving a set of linear equations. For example, does T affect f_7? We recall that f_7 is the equation

$$y_{1,1} - K_{1,1}x_{1,1} = 0$$

We of course know that T affects the $K_{1,1}$ calculation which in turn affects f_7. To prove this automatically, we could perform a logical Gaussian elimination on the equations and variables from T to f_7 on our list. (See figure 3.28.) If we assign T the role of tear variable and $K_{1,1}$ is

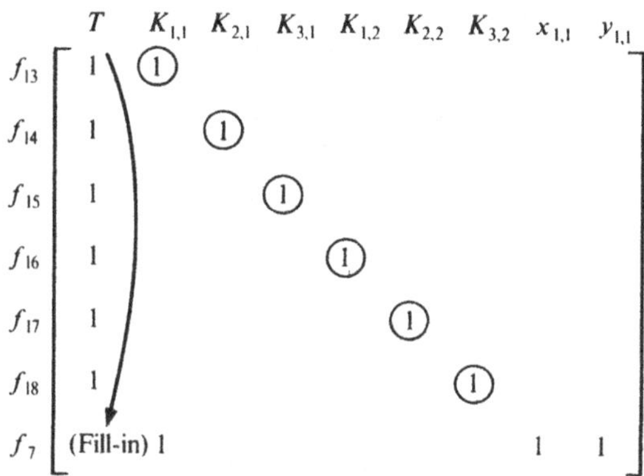

Figure 3.28. A 'logical' Gaussian elimination to see if f_7 is implicitly affected by T through equations f_{13} to f_{18}.

to be calculated from f_{13}, then eliminating $K_{1,1}$ from f_7 would cause the indicated fill-in. Thus T would appear implicitly in f_7. We note that if f_{13} were used to calculate T given $K_{1,1}$, then $K_{1,1}$ affects f_7 explicitly but T does not affect f_7 at all.

We can, for all assigned outputs, perform a logical Gaussian elimination getting the fill-in indicated by crosses in figure 3.29. The order in which to proceed is that given in our list shown in figure 3.27. The output assigned rows and columns can then be eliminated giving figure 3.30. Note that the explicit occurrences are indicated by 1s and the implicit occurrences by crosses. We now wish to assign the tear and decision variables to our best advantage. The goal is to make a number of smaller subproblems rather than one giant iteration loop, if possible.

The problem appears to have a break between equations f_9 and f_{10}. From f_{10} to the end we have six equations and six new variables $y_{1,2}$ to V.

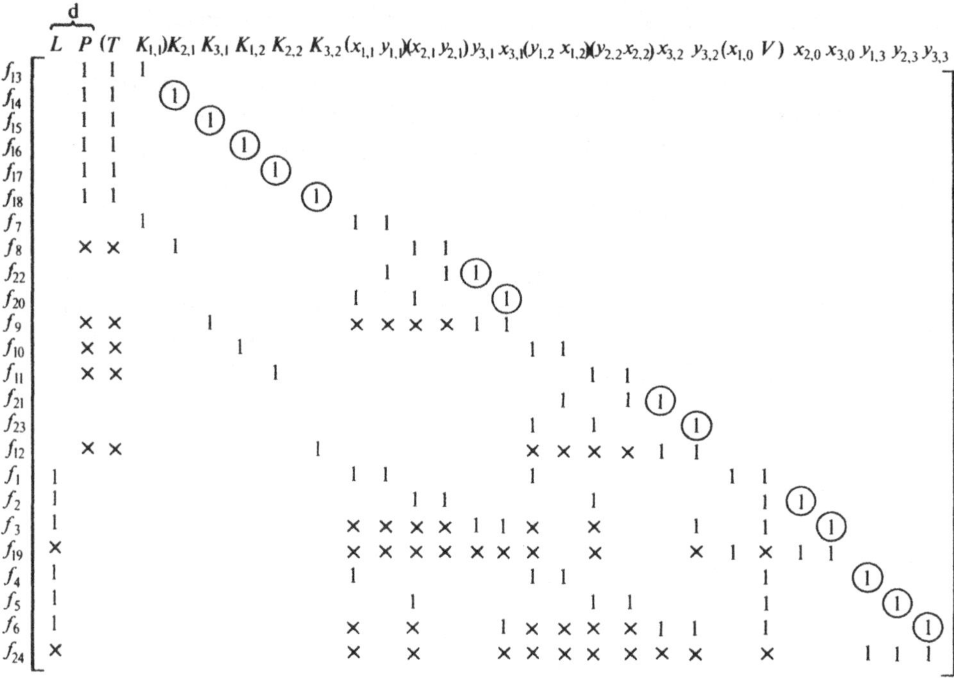

Figure 3.29. Fill-in of incidences from logical Gaussian elimination of assigned variables. d, decision variables.

Figure 3.30. Discovery of problem partitioning and first loop.

	(T $K_{1,1}$) ($x_{1,1}$ $y_{1,1}$)($x_{2,1}$ $y_{2,1}$)($y_{1,2}$ $x_{1,2}$)($y_{2,2}$ $x_{2,2}$)($x_{1,0}$ V)	Immediately follows	New variables
f_{13}	1 ①	T, $K_{1,1}$	2
f_7	v 1 ① 1	$x_{1,1}$, $y_{1,1}$	2
f_8	× ① 1	$x_{2,1}$, $y_{2,1}$	2
f_9	× × × × ⊗		0
f_{10}	× 1 1	$y_{1,2}$, $x_{1,2}$	2
f_{11}	× 1 1	$y_{2,2}$, $x_{2,2}$	2
f_{12}	× × × × ×		0
f_1	v 1 1 1 1 1	$x_{1,0}$, V	2
f_{19}	v × × × × × × 1 ×		0
f_{24}	v × v × × × × × ×		0

Therefore we shall attempt to partition the problem at this point. The last six equations will be used to calculate the last six variables. The first four equations involve six new variables; we shall select two of the variables from this set to be our two decision variables.

Proceeding, we assign $K_{1,1}$ to f_{13} and perform a logical Gaussian elimination. We assign $x_{1,1}$ to f_7, again performing a logical Gaussian elimination. Each resulting fill-in is shown in figure 3.30 with a 'v'. We assign $x_{2,1}$ to f_8 and are left to make an implicit assignment for f_9 from among the variables T, $y_{1,1}$ and $y_{2,1}$. Any of these will force a computational loop, so we choose $y_{2,1}$ since the loop will start just before calculating $x_{2,1}$, the smallest loop we can make. These assignments are shown in figure 3.30, with the computation loop shown in the inner square brackets. The variables T and $y_{1,1}$ will be the decision variables. The last six equations can be assigned as follows: $(f_{10}, y_{1,2})$, $(f_{11}, y_{2,2})$, $(f_{12}, x_{2,2})$, (f_1, V), $(f_{19}, x_{1,0})$ and $(f_{24}, x_{1,2})$. Figure 3.31 shows this assignment with each fill-in resulting from the performance of a logical Gaussian elimination indicated by a 'v'.

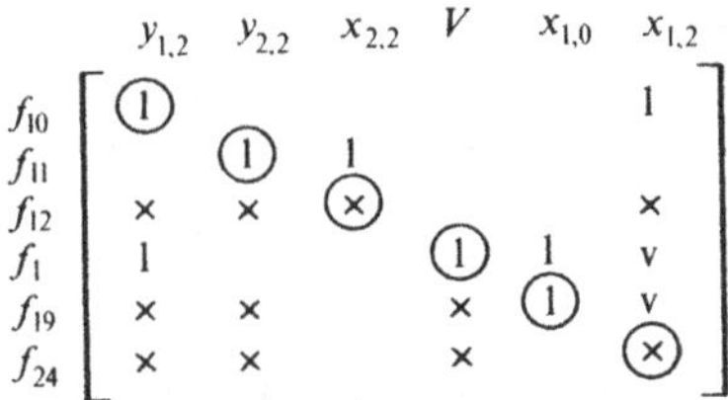

Figure 3.31. Assigning output variables in second major partition for problem.

Figure 3.32 indicates the final computational order. To construct figure 3.32, the equation order of our list in figure 3.27 is maintained. The decision variables are listed first and then the assigned output variables are listed in the order of the equations. We indicate two iteration loops: f_8 to f_9 with $y_{2,1}$ being the iterated tear variable, and f_{10} to f_{24} with $x_{2,2}$, $x_{1,0}$ and $x_{1,2}$ being three tear variables. The second loop could be solved at two levels as follows:

1. Guess $x_{1,2}$
2. Guess $x_{2,2}$
3. Use f_{11}, f_{21} and f_{23} to obtain $y_{2,2}$, $x_{3,2}$ and $y_{3,2}$ respectively
4. Evaluate f_{12} as an error function. If not essentially zero reguess $x_{2,2}$ and iterate from step 3
5. Solve f_1 to f_{19} by tearing and iterating $x_{1,0}$
6. Solve f_4 to f_6 for $y_{1,3}$, $y_{2,3}$ and $y_{3,3}$
7. Evaluate f_{24}. If not essentially zero, regress $x_{1,2}$ and iterate from step 2

It would probably be better, however, to solve it by guessing all three variables and converging the three functions f_{12}, f_{19} and f_{24} simultaneously using the methods of section 3.3.2. Note that two functions, f_{12} and f_{24}, are implicit in their tear variables; f_{19} is explicit.

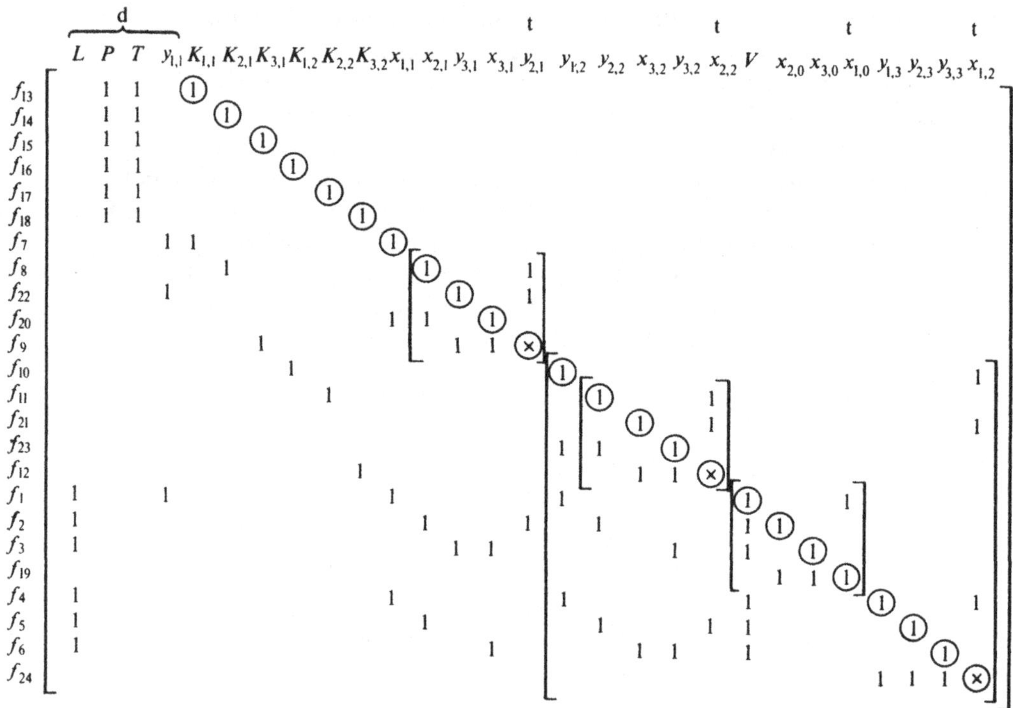

Figure 3.32. Final equation and variable ordering. Computational loops are illustrated. d, decision variables; t, tear variables.

The example just demonstrated illustrates that finding an effective solution procedure is a large effort; it also appears that we could automate it so that a computer could itself develop a solution procedure. We should note that all decisions made when working through the example were the apparent best we could make at that point in the problem, but we have no guarantee that better solution procedures could not be found. Christensen (1970) published a branch-and-bound algorithm which will guarantee the fewest tear variables. Whether, for a large problem, one can afford the search which results is still an open question. The algorithm described in this section is implemented and seems very effective for small problems with 50 to 100 equations (Westerberg and Shah, 1978).

We could use the above approach to select only the decision variables. Once selected, the problem is square, and we might be tempted then to reanalyse the problem as a square one, partitioning it, etc. Again the effort may not be worthwhile, in general, but this result has not been proved. It is clear that one may discover a partitioning not discovered here.

For solving a set of n linear equations involving m ($>n$) variables, one could use the sparse matrix strategies to find a pivot sequence. The variables not pivoted will become the decision variables. Again, once the decision variables are selected, the problem can be treated as square; partitioning can then be attempted.

The purpose of this section has been to demonstrate the approach one might take for nonsquare problems where the number of variables exceeds the number of equations. The ideas involved underlie much of the literature in this area.

3.4.3 Solving redundant equation sets

The other manner in which equation sets are not square is if the number of equations n exceeds the number of variables m. This case is one of overspecification since, in general, not all the equations can be satisfied at the same time. Two approaches are possible: (1) find a compromise solution, as is done in a least-squares formulation, or (2) satisfy m of the equations only and simply evaluate the remaining $n - m$. This latter approach was the one that we required earlier when we chose a tear variable in the isothermal flash example in section 3.4.1 and left a nonsquare system of equations. If, in fact, the equations are simply redundant, that is, they are all valid but one or more can be derived directly from the others, the least-squares error will be zero, indicating that all the equations are satisfied or the $n - m$ 'evaluated only' equations will have a zero value if the second approach is taken. A danger lurks for the second approach: one cannot normally locate the redundant equations easily unless the equations are linear. The redundant equations may be included in the set of m equations we propose to solve, making them unsolvable.

4

Physical property service facilities

A flowsheeting program almost always contains a physical property 'service', since the quality of process design is ultimately dependent on the way in which the laws of physics and chemistry are applied to the problem. Accordingly, the quality of this service is an important consideration to the user of a flowsheeting system. The provision of these facilities is, however, a technology in its own right, and in this book we shall only look at overall qualitative characteristics of physical property usage in the flowsheeting context.

The physical property service has to perform a number of tasks, but the most useful of these are:

(1)   to supply estimates repetitively for a number of different physical properties while the simulation is in execution.

(2)   to provide the user with values of properties of interest during the calculation and/or at simulation completion, for subsequent use in other calculations.

(3)   to allow the user to input his own special data for new components and transform it into the form required by the system during a simulation.

(4)   to supply the user with a means to estimate properties where little, except perhaps chemical structure, is known about a particular chemical compound. Again the system must put these estimates into the form needed by the simulation during execution.

4.1 The data cycle

It is perhaps useful to review the flow of physical property data from source to use in flowsheeting. Figure 4.1 illustrates this flow. The new data originate from measurement in the laboratory and are often published in the open literature. To start a physical property calculation system, both sources are used to build up what one might term a library of raw data. The bulk of data is for pure components; some is for mixtures of usually no more than two or three components over limited

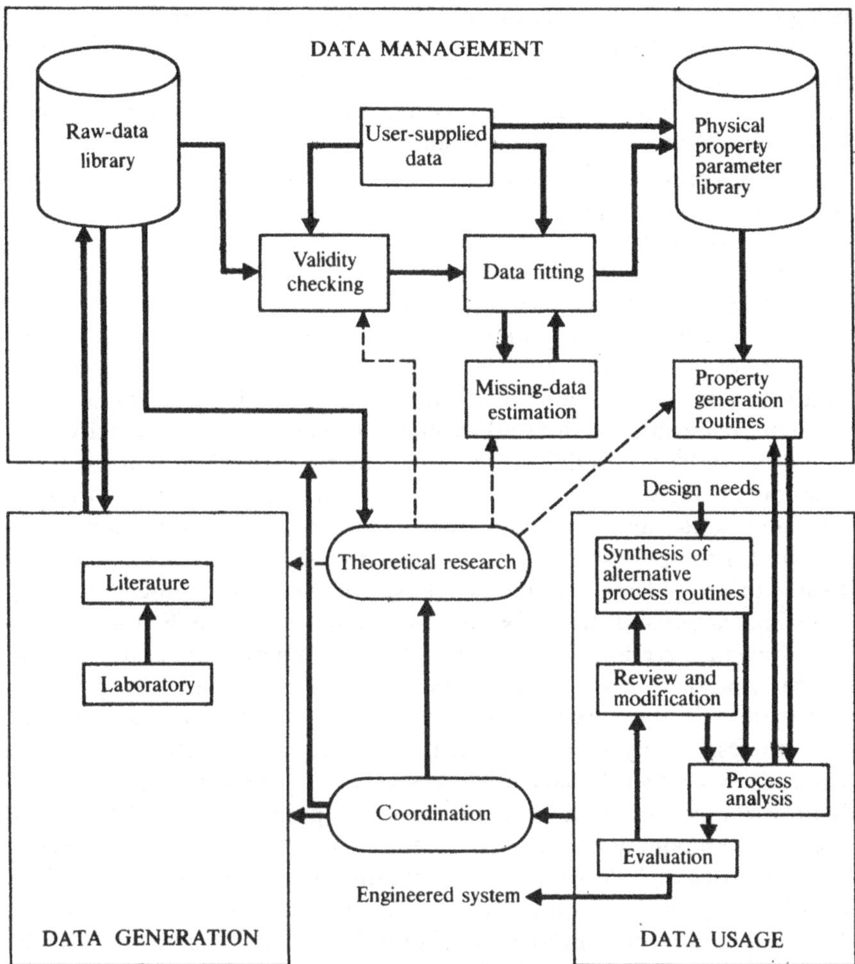

Figure 4.1. The generation, management and use of physical property data in engineering design.

ranges of temperature and pressure. These data must be evaluated, often with some being rejected. Validity checking is a combination of consistency checking and subjective evaluation of the data source. Some laboratories and some people are assessed as being experimentally superior to others at producing data, so conflicts in values must be considered against this assessment.

Most laboratory data are presented in tables and must be condensed to a form more readily usable. Also conflicting data, where none can be rejected out of hand, require that a compromise set of values be produced. The resulting evaluated data are reduced to equation form using some means of statistical fitting, such as least-squares, to establish a set of parameters for use in the equation. The form of such equations is often determined from theoretical consideration, and, of course, must be if data are to be extrapolated with any confidence beyond the range where experimental values are available.

The above considerations apply to data on the pure component, but the bulk of physical properties required for design are for mixtures of components appearing together in one, two and sometimes more phases. Few experimental data exist for mixtures, relatively speaking, and virtually all design needs are for mixture properties. Therefore these properties must be estimated, and considerable theoretical effort is devoted to this one goal in order to enable reasonably accurate prediction of the behaviour of the mixtures. The difficulties are evident enough. For example, how does one estimate the heat capacity of a liquid mixture? One might be tempted to mole-fraction average pure-component liquid heat capacities, but why not weight-fraction or volume-fraction average? And what if one of the components is in the liquid at a temperature well above its critical temperature? No liquid heat capacity exists for it. To delve further into these intriguing questions is beyond the scope of this chapter, so we shall stop.

In view of this enormous theoretical effort to permit accurate physical property prediction for mixtures, the models used for data fitting and for calculation of the mixture properties are in rather a state of flux. Nonetheless, many models do exist and are used fairly successfully, particularly on well-behaved families of components such as nonpolar hydrocarbons. The models require a set of parameters to characterize the pure components; this set normally consists of fitted parameters of the type discussed above, plus basic thermodynamic constants, and may also include additional data that characterize the interactions between binary mixtures of components. Examples of basic constants are the critical properties: T_c, P_c and Z_c (critical temperature, pressure and compressibility factor, respectively). Also many data are correlated against the acentric factor ω which is an indicator of the nonspherical nature of a molecule's force field.

Fitted parameters include such items as heat capacity curves reduced to polynomial form against temperature. Other data are for the transport properties such as viscosity, thermal conductivity and diffusivity; these also are normally in polynomial form. Thus the data-fitting step is to convert raw data to a compact form suitable for use in the property generation routines that embody the theoretical models.

In some instances there are little or no 'raw' data on which to base the creation of these parameters. Estimation of missing data is normally a 'roadmapping' problem. Many methods usually exist for estimating a particular physical property and they often use different information about the compound as a starting point for the calculation. To find an estimate of physical property *a* may require knowledge of estimates of properties *b* and *c* by one estimation procedure and knowledge of properties *c* and *d* by another. These in turn, if estimated, will require other physical properties. The estimation errors propagate through such a series of calculations, and one can keep track of their approximate magnitudes. This error estimate can give one a means, though not a decisive one, to choose among the methods used. One may prefer consistency with the estimation of other properties, and choose a path considered less accurate because the other properties use much of the same path and are themselves more accurate this way. The least amount of data required for some systems is the chemical structure of the compound. From this starting point critical properties are estimated, then the vapour pressure curves and so forth. These types of estimations may permit one to do a design and then assess whether any additional accuracy is needed.

The final aspect of the data cycle is, of course, its use in the design step. This use occurs principally in the analysis phase where flowsheeting occurs. It is at this stage where the quality of the data is really assessed, since it is only useful if it enables the engineer to fulfil the design needs. This is also the stage that acts as a springboard to stimulate the demand for new data and better methods, so completing the data cycle.

4.2 Computerized physical property systems

Flowsheeting calculations tend to have a voracious appetite for physical property estimations. To model a distillation column one may request estimates for chemical potential (or fugacity) and for enthalpies 10 000 (but usually many more) times. Depending on the efficiency of the programming, these calculations could represent 80% or more of the computer time taken to do a simulation. With this requirement to handle an enormous number of requests, the design of the physical property estimation system must be done with extreme care. All conceivable preprocessing which might be done has to be done. It is much too costly during a simulation to have a request for the heat capacity of a mixture trigger off a calculation reaching all the way back to the raw-data library. The steps leading to the physical property parameter library must, if at all possible, be done once and for all before the simulation, and the point generation routines must be able to find all the data they need in this library. For added efficiency the data for the chemical species actually in the process should be extracted into a smaller data library for use during the simulation.

Figure 4.2 summarizes the general features of what one may expect to

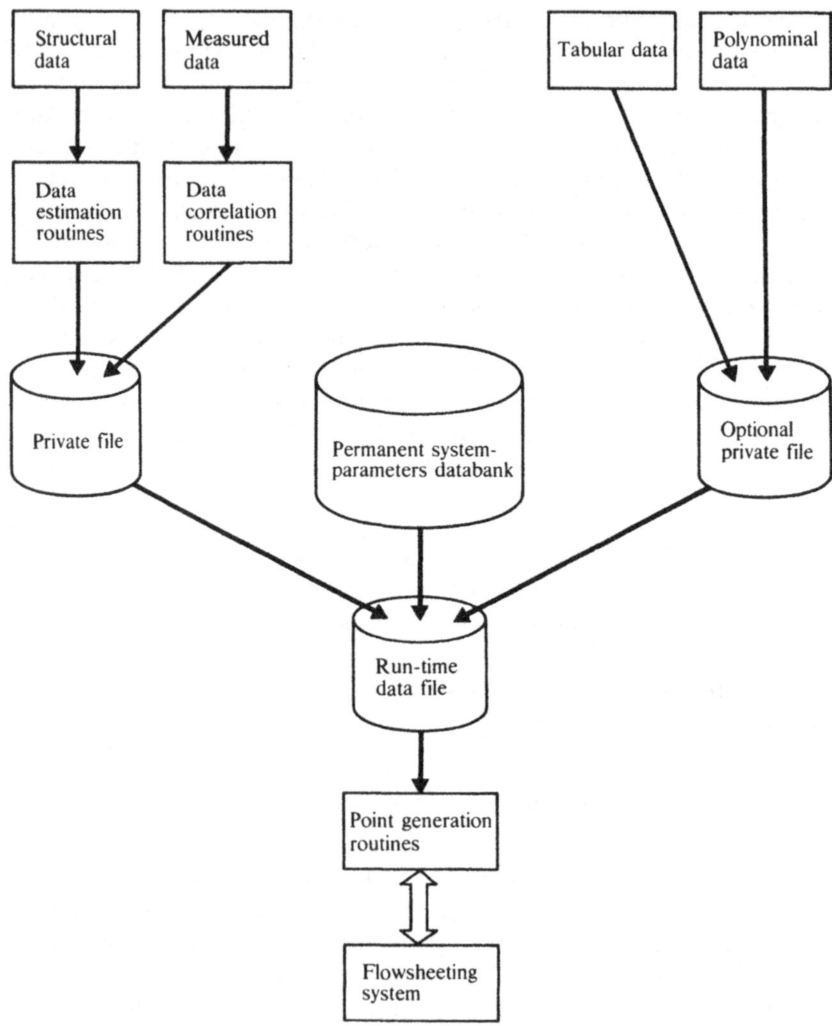

Figure 4.2. General features of a computerized physical property system.

be supplied in the type of physical property data service provided for use with a flowsheeting system. First consider the central column of the diagram, which represents the basic set of facilities.

The simulation models of the flowsheeting system need to make frequent requests for properties at specific temperatures, pressures and compositions. These calls for data are usually made in a rigorously defined manner which is independent of both the point data generation models and the particular components. These point generation routines provide the property values using the selected methods that base their calculations on a set of parameters for each component, as discussed in the previous section.

At run time it is desirable to store only the parameters for those particular components involved in the simulation, and so this set of basic data is normally copied over from the permanent system parameters databank into run-time data locations. In typical flowsheeting problems this involves collection of particular parameters for 5–20 components from a databank with extensive sets of data for 250–750 compounds.

Use of such a data system is very easy, provided that the permanent system databank has entries for the components of interest. Thus it is often of importance to the user that the databank is extensive, and not restricted to a small class of compounds or to a particular set of data generation methods. Unfortunately, the number of chemical species is enormous and expanding at a rapid rate. Accordingly, for wide application it is necessary that the physical property data service of the flowsheeting system embodies facilities for the utilisation of user-supplied data.

From the systems point of view the easiest way to accommodate this requirement is for the user to be able to create his own equivalent of the permanent system databank by explicitly entering data in the same format. This option is easily made available and allows the user to build up his own private databank for use independently or in conjunction with the system databank.

Data supply in this way normally requires that the user has (or has access to) expertise in physical property data and possibly also of computer usage. In these circumstances he may also wish to provide data in tabular or polynomial form for use by an appropriate set of interpolative point generation routines. This facility is shown at the top right of figure 4.2.

In addition to these facilities for supply of data in an explicit form for direct usage by the system, there are also those options designed for the calculation of the parameters used by the system's point generation routines. Two obvious categories of this type can be identified and are included at the top left of figure 4.2. The first of these applies to the correlation of raw data and is most commonly applied to the estimation of binary interaction parameters.

The second category differs from those discussed above since it relates, in the main, to those situations for which no data or only characterizing data exist. In such cases this small set of characterizing data, or in its absence, structure data, are utilised to estimate a set of parameters of the type required by the point generation routines. One notable specific example of this type of facility is the creation of data sets for petroleum boiling fractions from information on average boiling point and density.

All the above facilities form part of the spectrum of options that, in addition to the permanent system databank, enable the engineer to get the most out of a flowsheeting system. If we consider these as components of a whole system, figure 4.3 illustrates a way by which the various options together could be used to provide weightings for creation and updating of the parameters databank.

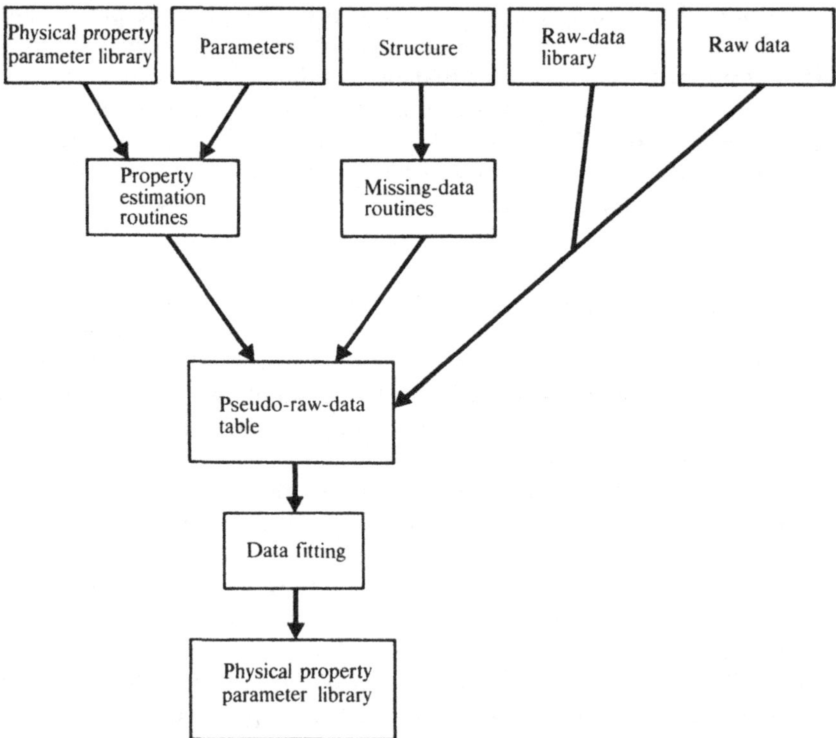

Figure 4.3. Possible options to supply physical property data for a compound. A 'weighted' combination of all the sources could, in principle, be used.

A general physical property service of this type could allow input from any or all of the sources for physical property data. The service will, if needed, merge all of these inputs to get a best set of data for the physical property parameter library, possibly by predicting the properties resulting from each input and putting them into a table of pseudo raw data. Then these generated data can be fitted as if they were direct user input. The data sources can be weighted so that the user can emphasize those sources that he feels are more reliable.

Having discussed the overview of a physical property service, the next section will discuss rather briefly the type of property calculations required by a flowsheeting system.

4.3 Physical property calculations

This section will discuss briefly a typical set of physical properties which might be provided with a flowsheeting package. Table 4.1 lists many, if not most, of these.

Table 4.1. *More-common physical properties used in design for single-phase mixtures. Input data are P, T, z_i for all i and phase indication (vapour or liquid)*

Thermodynamic	Transport
Equation of state	Viscosity
*Density	Thermal conductivity
Compressibility	Diffusion coefficients
Thermal expansion coefficient	
Heat capacity	
*Enthalpy	
Heat of mixing	
*Entropy	
Gibbs free energy	
Heats of formation	
Entropy of formation	
Heat of reaction	
Free energy of reaction	
*K-values	
Activity coefficients	
Fugacity coefficients	
Chemical potential	
Fugacity	
Surface Tension	

* Properties almost always available to user of a flowsheeting program.

The asterisks indicate those properties quite commonly available to the engineer who may be writing his own module to model a special unit for his plant simulation. The remaining are less commonly provided.

A substantial amount of the calculations for a flowsheeting package involves two-phase systems of mixtures. Figure 4.4 aids in defining the nomenclature for table 4.2, which gives a number of the vapour–liquid equilibrium services commonly provided. The composition of the mixture is characterized overall by mole fractions z_i and for each of its two phases by compositions y_i and x_i. Also its total amount of F, with V being the number of moles in the vapour phase, and L the number of moles in the liquid phase. We can view it as an equilibrium flash unit placed in a stream, splitting the stream into two flows and then remixing.

We note that the value of F, L and V are not relevant; only their relative values, characterized by the vapour fraction $\varphi = V/F$, are needed.

Returning to table 4.1, we have partitioned the physical properties into thermodynamic properties and transport properties, which characterize equilibrium situations and rate situations. The thermodynamic properties are grouped roughly to reflect their common origin or

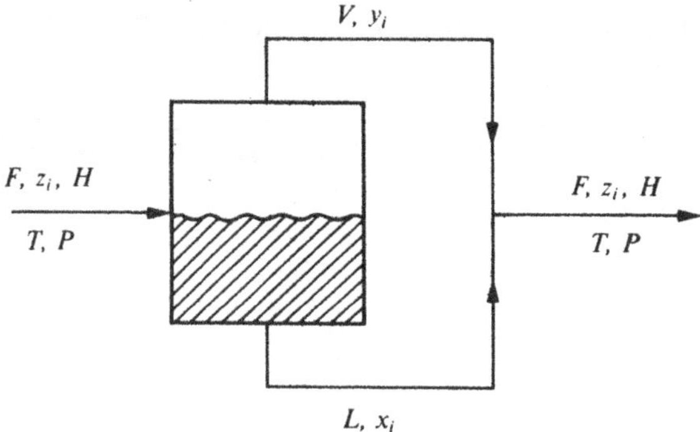

Figure 4.4. Two-phase model for a mixture. F, V, L: amounts (mol); c: number of components in the mixture; x_i, y_i z_i $i = 1, 2, \ldots, c$: mole fractions; T: temperature (K); P: pressure (bar); $\varphi = V/F$: vapour fraction ($= 1$ all vapour, $= 0$ all liquid); H: molar enthalpy of mixture (J/mol).

Table 4.2. *Vapour–liquid equilibrium properties generally available*

Property	Input	Calculated
K-values	T, P, x_i, y_i, all i	Equilibrium ratios K_i, all i
Dew point	P, z_i, all i, $V/F = 1$	T
Bubble point	P, z_i, all i, $V/F = 0$	T
Dew point pressure	T, z_i, all i, $V/F = 1$	P
Bubble point pressure	T, z_i, all i, $V/F = 0$	P
Heat of vaporization	P, z_i, all i	H (dew point)$-H$ (bubble point)
Isothermal flash	T, P, z_i, all i	$V/F, Q, x_i$ and y_i (all i)
Adiabatic flash	H, P, z_i, all i	V/F and T, x_i and y_i (all i)
General flash	Any two of $T, P, H, V/F$, and z_i, all i	Other two of $T, H, V/F$, x_i and y_i (all i)
Isothermal three-phase flash	T, P, z_i	Number of phases, compositions of each, H of each, phase fraction of each

purpose. The first group are PVT-relationships, and the second are for relating the enthalpy and entropy of a mixture to a standard state for the pure components. The properties of the third grouping allow one to relate the standard states of pure components to one another and are needed only if chemical reaction takes place. Vapour–liquid equilibria are commonly characterized by using the properties of the fourth set. Surface tension is a property we seem forced to list by itself; it is not a bulk property, but rather an interfacial property, and is often required in design correlations.

We might note that many additional and useful calculations are available if the physical properties listed are available on request. A single-phase mixture is characterized by its temperature T, pressure P, phase indicator φ (= 0 or 1) and its compositions z_i, i = 1, 2,. . ., c. We may wish to supply the enthalpy H, P, φ and z_i, and find the associated temperature (for example when solving the equations for a heat exchanger). The physical property package will allow us to solve

$$H = F_H(T, P, z_i \text{ (all } i), \varphi)$$

directly. By writing

$$g(T) = H - F_H(T, P, z_i \text{ (all } i), \varphi) = 0$$

we have an implicit function in T which can be solved using a root-finding technique, such as the bounded secant method described in section 3.1. If this is provided as a routine, it becomes a useful part of the physical property package. Many other general routines are also useful: (1) find P given entropy S, T, φ and z_i (all i); (2) find T given density, P, φ and z_i (all i).

The two-phase calculations are not as straightforward as so many difficulties can arise. For example, what if the conditions given as input result in a single-phase rather than a two-phase mixture? This situation has to be trapped as it frequently occurs. The isothermal flash is easy to trap as one can calculate the dew point temperature to see if the specified temperature is above it and the mixture is then all vapour. If not, one can compare to the bubble point temperature and, if below that, the mixture is all liquid. Otherwise it must be two phase.

The adiabatic flash calculation is similarly checked. Here one is adjusting both the temperature T and the vapour fraction φ to satisfy the heat balance and the equilibrium relationships. If the heat balance indicates that the mixture must be hotter than it is at its dew point, then it must be all vapour. Similarly, if the heat balance at the bubble point indicates that it must be colder still, the mixture must be all liquid. Otherwise, it is again two phase.

The last flash computation which might be included in the physical property service is that for a three-phase flash – two liquid and one vapour phase. This service means that the vapour–liquid equilibrium (VLE) models used to predict the chemical potentials of each species in each phase must also predict the existence of more than one liquid phase. The Gibbs free energy is at a minimum at equilibrium so the model must predict a lower Gibbs free energy for three phases than for two, and this comparison must be made to allow one to determine the number of phases.

As an exercise the reader might devise solutions to the following problems. Be sure to consider the possible difficulties which might arise. Do not worry about three phases occurring.

(1)   Find the temperature of a mixture given H, P and z_i (all i). The phase of the mixture is not known.
(2)   Three streams for a countercurrent heat exchanger are known (i.e. their T, P, φ, H and z_i (all i) values are known). Find the conditions of the remaining stream. Will the heat exchanger work?
(3)   For a compressor, the inlet stream conditions are known and the desired outlet pressure is specified. The compressor is isentropic. Discuss how to calculate the outlet stream conditions.

We shall now ask some questions related to these problems.

Question relating to problem 1. Did you use the adiabatic flash routine? It does precisely this calculation.

Questions relating to problem 2.   A heat balance on the three streams provides the enthalpy of the remaining stream. If pressure drops are ignored, the outlet pressure is available, and we again require the services of the adiabatic flash calculation. If pressure drops are not ignored, the calculation could get messy. We shall skip this problem here. Of interest, did you consider that the temperatures might have crossed over, making the design impossible? If either stream crosses either its dew or bubble point within the exchanger, a crossover may occur, and this may not be made apparent by examining the inlet and exit temperatures. How did you check for this? (You never thought of it?)

Questions relating to problem 3.   The previous problem should serve as a warning. Did you check the feed to see that it was all vapour? Compressors self-destruct if the inlet stream is two phase. The simplest approach here, given an all vapour feed, is to note that the entropy S of the inlet equals the entropy of the outlet. Thus we need to find T implicity by finding a root to

$$g_S(T) = S - F_S(T, P_{\text{out}}, z_i \text{ (all } i), \varphi = 1) = 0$$

where $F_S(T, P, z_i \text{ (all } i), \varphi)$ is available as a calculation in the physical properties package.

It is beyond the intended scope of this book to delve further into the calculation of physical properties. The attempt here has been to give an understanding of their importance and how they relate to process flowsheeting. Much useful information on this subject is contained in the book by Reid, Prausnitz and Sherwood (1977).

5

Degrees of freedom in a flowsheet

We have continually mentioned the degrees of freedom for a flowsheet. In fact a very difficult problem in flowsheeting is to specify the problem so that the flowsheeting program solves the desired problem. Tied up with this problem is one of correctly identifying the number of degrees of freedom and finding a legitimate means to satisfy them. In this chapter we shall investigate this problem in some detail, mostly by specific example. We shall first give the most basic view of the number of degrees of freedom, then establish how many must exist for a process stream. Next we shall consider the number which may occur for a unit model, illustrating with several simple unit types: a mixer, a flash unit, an extent of conversion reactor, a simple flow splitter and pressure changing devices. With these models we will build a small flowsheet comprising them and examine its degrees of freedom.

5.1 Degrees of freedom

The simplest definition has already been given for the number of degrees of freedom of a problem involving the solution of a set of equations. If our problem is one with n independent equations in m ($m > n$) variables, then the number of degrees of freedom d is given as

$$d = m - n \tag{5.1}$$

This formula is based on the idea that n well-behaved independent equations can be used to solve for exactly n variables. Thus $m - n$ of the variables must receive their values from other means before we can solve the problem. Whatever the other means, they must be equivalent to writing an additional set of $m - n$ independent equations.

We can consider the classical rule for degrees of freedom, the phase rule. The phase rule says that the number of degrees of freedom for a multiphase system, where all phases are in equilibrium, is

$$d = c - p + 2 \tag{5.2}$$

where c is the number of components and p the number of phases.

This rule applies only for the intensive variables of the system and says nothing about the extensive variables. Intensive variables are those which are quantity independent, such as T, P, mole fraction and chemical potential. Extensive variables are quantity dependent, doubling in value if the size of the system doubles, and include total volume, total mass and total enthalpy. It is not too difficult to prove this rule using our definition for the degrees of freedom. The equations modelling a multiphase system are

Number of
equations

Thermal equilibrium:

$$T_1 = T_2 = \ldots = T_p \equiv T \qquad\qquad p \qquad (5.3)$$

Pressure equilibrium:

$$P_1 = P_2 = \ldots = P_p \equiv P \qquad\qquad p \qquad (5.4)$$

Chemical potential:

$$\mu_{1j} = \mu_{2j} = \ldots = \mu_{pj} \equiv \mu_j \qquad\qquad cp \qquad (5.5)$$

Relating chemical potential to measurable quantities:

$$\mu_{kj} = \mu_{kj}(T, P, x_{kq}; q = 1, 2, \ldots, c) \qquad cp \qquad (5.6)$$
$$k = 1, 2, \ldots, p \qquad j = 1, 2, \ldots, c$$

Mole fraction sum equations:

$$x_{k1} + x_{k2} + \ldots + x_{kc} = 1 \qquad k = 1, 2, \ldots, p \qquad\qquad p \qquad (5.7)$$

Total $n = 3p + 2cp$ equations

For each phase we have the variables T_k, P_k, μ_{kj}, x_{kj}; $j = 1, 2, \ldots, c$. This gives $2c + 2$ variables per phase plus $c + 2$ extra 'overall' variables $(T, P, \mu_j; j = 1, 2, \ldots, c)$ for a total of

$$m = (2c + 2)\, p + c + 2$$

Applying equation (5.1) then gives the phase rule:

$$d = m - n = c - p + 2.$$

While the principle behind equation (5.1) is straightforward, the application of it is far from simple. It is not an easy task to write the set of equations which model a unit operation even if the theory underlying the equations is not a problem. One often writes too many equations, not realizing that one or more of them can be derived algebraically from the others. An example is for the material balance equations of a unit. Figure 5.1 illustrates a unit with two inputs and one output (a mixing unit for example).

We can write the material balance equations

$$x_{j,1}F_1 + x_{j,2}F_2 = x_{j,3}F_3 \qquad j = 1, 2, \ldots, c \qquad (5.8)$$

$$F_1 + F_2 = F_3 \qquad (5.9)$$

plus the mole fraction sum equations

$$\Sigma x_{j,1} = 1 \qquad (5.10)$$

$$\Sigma x_{j,2} = 1 \qquad (5.11)$$

$$\Sigma x_{j,3} = 1 \qquad (5.12)$$

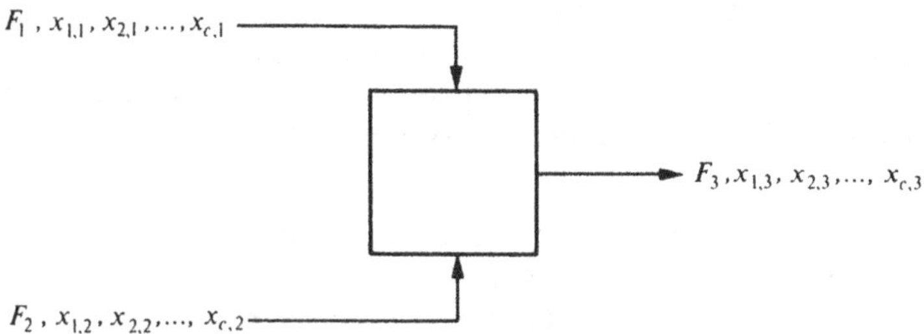

Figure 5.1. A simple unit.

If we do this, we have written one equation too many. Equation (5.12) can be derived from the others by the following. Sum equations (5.8) over all j:

$$(\Sigma x_{j,1})F_1 + (\Sigma x_{j,2})F_2 = (\Sigma x_{j,3})F_3 \qquad (5.13)$$

Then substituting equations (5.10) and (5.11) into equation (5.13) gives

$$F_1 + F_2 = (\Sigma x_{j,3})F_3 \qquad (5.14)$$

and this combined with equation (5.9) gives us equation (5.12).

For a flowsheet the problem can be very difficult and one's intuition and/or logic may easily fail. If you are a chemical engineer, it is only necessary to recall your stoichiometry course to recognize the truth of this observation.

5.2 Independent stream variables

What is the minimum number of variables required to specify fully a stream? We define a stream as the flow of material between two units in a flowsheet. The variables we normally associate with a stream are its temperature, pressure, total flow, overall mole fractions, phase fractions and phase mole fractions, total enthalpy, phase enthalpies and so forth. How many of these are needed to fix completely the stream if we assume phase and chemical equilibrium exists?

Intuition gives us the correct response here. We know, without counting equations and variables, that, if we specify the temperature, pressure and individual component flows, the stream must be fully specified. We know this must be true because, if we create such a stream in the laboratory and fix its temperature, its pressure and its component flows, the stream will break into phases (if it is multiphased at equilibrium) and so forth. (We could get into difficulty with temperature for pure components. If for a pure component we fix the pressure and the individual component flows, and then we choose the temperature to be the boiling point temperature, we are at a loss to state the vapour fraction for it. In this case one can replace temperature with enthalpy, and then one has no such problem.) The question now arises as to whether we can prove our intuitive response. We can proceed by writing the equations needed to model our system.

Duhem's theorem (Prigogine and Defay, 1954) provides us with a needed result, and it is proved in just this way. The theorem states: 'Whatever the number of phases, of components or of chemical reactions, the equilibrium state of a closed system, for which we know the total initial masses of each component, is completely determined by two independent variables.' Can we classify a stream as a closed system? We can do so by taking a fixed amount of it (say its flow for one second) and trapping it in a container. Because we know the compositions and total flow of the stream, we know the total initial masses we have trapped. The theorem says we need only fix two additional independent variables, such as temperature (or enthalpy to be safe) and pressure, to fix the equilibrium state of the system.

The proof follows. We shall do it for a particular system to reduce the chance of getting lost in the notation. Phase equilibrium implies that the equations which we wrote in the previous section to prove the phase rule hold true. These equations are equation (5.3) to (5.7). Added to these are material balance equations and reaction equilibrium equations.

The example system we shall consider will involve the $c = 5$ species H_2, H, Br_2, Br and HBr related by the $r = 3$ independent reactions.

$$2H = H_2$$
$$2Br = Br_2$$
$$H_2 + Br_2 = 2HBr$$

It is convenient to rewrite these reactions in the form

$$\text{Reaction } i: \sum_{j=1}^{c} v_{ij} M_j = 0 \quad i = 1, 2, \ldots, r \qquad (5.15)$$

where v_{ij} are stoichiometric coefficients, M_j are species names and r is the number of *independent* reactions. We will have v_{ij} positive for products of a reaction (species on the right hand side of the chemical reaction as written) and v_{ij} negative for reactants. For our three reactions we get

$$H_2 - 2H = 0$$

$$Br_2 - 2Br = 0$$

$$2HBr - H_2 - Br_2 = 0$$

For reaction 1, equation (5.15) fully expanded is

$$1H_2 - 2H + 0Br_2 + 0Br + 0HBr = 0$$

so $v_{11} = 1$, $v_{12} = -2$, $v_{13} = 0$, and so forth.

We next introduce the concept of 'extent of reaction', denoted ξ_i for reaction i. We say each reaction i occurs to extent ξ_i, and for our example this means $2\xi_1$ moles of H disappear to form ξ_1 moles of H_2 by reaction 1, $2\xi_2$ moles of Br disappear to form ξ_2 moles of Br_2 by reaction 2, and finally ξ_3 moles each of H_2 and Br_2 disappear to form $2\xi_3$ moles of HBr by reaction 3. An overall material balance on each species gives

$$
\begin{aligned}
n_{H_2} &= n^{o}_{H_2} + \xi_1 - \xi_3 \\
n_{H} &= n^{o}_{H} - 2\xi_1 \\
n_{Br_2} &= n^{o}_{Br_2} + \xi_2 - \xi_3 \\
n_{Br} &= n^{o}_{Br} - 2\xi_2 \\
n_{HBr} &= n^{o}_{HBr} + 2\xi_3
\end{aligned}
$$

where n^{o}_{j} is the original number of moles of species j present. These equations are of the form

$$n_j = n^{o}_j + \sum_{i=1}^{r} v_{ij}\, \xi_i \quad j = 1, 2, \ldots, c \tag{5.16}$$

Reaction equilibrium occurs when the Gibbs' free energy G is at a minimum. Any variation in G around the equilibrium point must therefore be zero, i.e.

$$dG = -SdT + VdP + \sum_{j=1}^{c} \mu_j dn_j = 0$$

If T and P are constant this reduces to

$$\sum_{j=1}^{c} \mu_j\, dn_j = 0 \tag{5.17}$$

However the dn_j are constrained to abide by the reaction stoichiometry and material balances as given by equations (5.16). We take variations of equation (5.16) for our example and get

$$
\begin{aligned}
dn_{H_2} &= d\xi_1 - d\xi_3 \\
dn_{H} &= -2d\xi_1 \\
dn_{Br_2} &= d\xi_2 - d\xi_3 \\
dn_{Br} &= -2d\xi_2 \\
dn_{HBr} &= 2d\xi_3
\end{aligned}
$$

In general then these are of the form

$$
dn_j = \sum_{i=1}^{r} v_{ij} d\xi_i \quad j = 1, 2, \ldots, c \tag{5.18}
$$

Substituting these into equation (5.17) gives for our example

$$
\begin{aligned}
\mu_{H_2}(d\xi_1 - d\xi_3) &+ \mu_{H}(-2d\xi_1) + \mu_{Br_2}(d\xi_2 - d\xi_3) \\
&+ \mu_{Br}(-2d\xi_2) + \mu_{HBr}(2d\xi_3) \\
= (\mu_{H_2} - 2\mu_{H})d\xi_1 &+ (\mu_{Br_2} - 2\mu_{Br})d\xi_2 \\
&+ (2\mu_{HBr} - \mu_{H_2} - \mu_{Br_2})d\xi_3 = 0
\end{aligned}
$$

Since the reactions are independent, each ξ_i can be varied independently. To be at equilibrium the coefficients for each must be zero, i.e.

$$
\begin{aligned}
\mu_{H_2} - 2\mu_{H} &= 0 \\
\mu_{Br_2} - 2\mu_{Br} &= 0 \\
2\mu_{HBr} - \mu_{H_2} - \mu_{Br_2} &= 0
\end{aligned}
$$

or in general

$$
\sum_{j=1}^{c} v_{ij}\mu_j = 0 \quad i = 1, 2, \ldots, r \tag{5.19}
$$

These look exactly like our reaction equations (5.15) with each chemical potential replacing its corresponding species name. We should note exactly what we did here. Equations (5.17) express the reaction equilibrium relationships we need; however, the variations in the number of moles of each of the $c = 5$ species are not independent. They must satisfy the reaction stoichiometry of the $r = 3$ independent reactions. Thus equations (5.17) do not give rise to $c = 5$ reaction equilibrium relations but rather $r = 3$ as we have just demonstrated.

The last equations we can write are the material balance relationships among the phases, and these relate phase mole fractions x_{kj} to the final total number of moles of each component. If each phase has m_k moles in it, we can write for each component.

$$
\sum_{k=1}^{p} x_{kj} m_k = n_j \tag{5.20}
$$

Equations (5.16) and (5.20) are the required material balance equations, and equations (5.19) the reaction equilibrium equations. Collecting them all together we have

Number of
equations

Thermal equilibrium:

$$T_1 = T_2 = \ldots = T_p \equiv T \qquad\qquad p$$

Pressure equilibrium:

$$P_1 = P_2 = \ldots = P_p \equiv P \qquad\qquad p$$

Chemical potential:

$$\mu_{1j} = \mu_{2j} = \ldots \mu_{pj} \equiv \mu_j \quad j = 1, 2, \ldots, c \qquad\qquad cp$$

Relating chemical potential to measurable quantities:

$$\mu_{kj} = \mu_{kj}(T, P, x_{kq}; \quad q = 1, 2, \ldots, c) \qquad\qquad cp$$
$$k = 1, 2, \ldots, p \quad j = 1, 2, \ldots, c$$

Mole fraction sum equations:

$$x_{k1} + x_{k2} + \ldots + x_{kc} = 1 \quad k = 1, 2, \ldots, p \qquad\qquad p$$

Material balance:

$$n_j = n_j^\circ + \sum_{i=1}^{r} v_{ij}\xi_i \qquad\qquad c$$

$$\sum_{k=1}^{p} x_{kj} m_k = n_j \quad j = 1, 2, \ldots, c \qquad\qquad c$$

Reaction equilibrium:

$$\sum_{j=1}^{c} v_{ij}\mu_{ij} = 0 \qquad i = 1, 2, \ldots, r \qquad\qquad r$$

Prespecified:

$$n_j^\circ = \text{given} \quad j = 1, 2, \ldots, c \qquad\qquad c$$

$$\text{Total} \quad n = (2c+3)p + 3c + r \text{ equations}$$

Number of
variables

For each phase, $k = 1, 2, \ldots, p$:

$$T_k, P_k, m_k, x_{kj}, \mu_{kj} \quad j = 1, 2, \ldots, c \qquad\qquad (2c+3)p$$

For each species, $j = 1, 2, \ldots, c$:

$$\mu_j, n_j, n_j^\circ \qquad\qquad 3c$$

Number of
variables

For each reaction, $i = 1, 2, \ldots, r$

$$\xi_i \qquad\qquad r$$

Overall:

$$T, P \qquad\qquad 2$$

$$m = (2c+3)p+3c+r+2 \text{ variables}$$

The number of degrees of freedom is then

$$d = m - n = 2$$

The total degrees of freedom associated with a stream is therefore 2 in addition to the molar flowrates for each component, whether reaction occurs and/or whether the stream splits into more than a single phase. The variables normally specified for a stream are P, T, total molar flow and $c-1$ overall mole fractions, or equivalently P, T and the individual overall component molar flowrates.

For a flowsheeting system, reaction is not permitted for a stream. If reaction is to occur, the unreacted stream becomes a feed to a reactor unit, where reaction is allowed, and the reacted feed (perhaps at equilibrium) becomes the reactor outlet stream.

5.3 Degrees of freedom for a unit

We shall consider several rather simple unit-type models here to establish how one ascertains the degrees of freedom associated with a unit. A related development appears in Rudd and Watson (1968).

Our first unit will be the simple mixer unit shown in figure 5.2. The equations we shall write for it are the following.

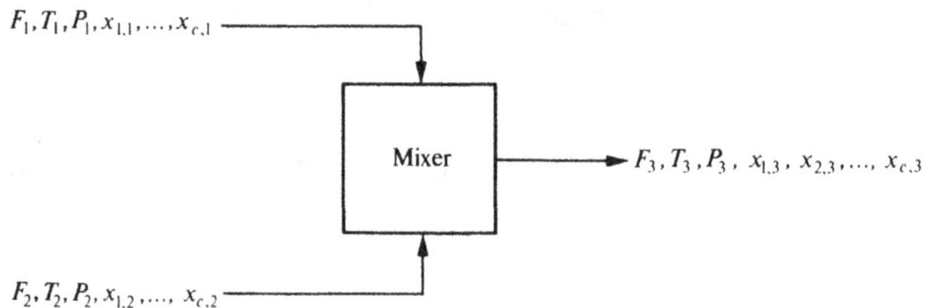

Figure 5.2 A simple mixer unit.

Material balance for each component:

$$x_{i,1}F_1 + x_{i,2}F_2 = x_{i,3}F_3 \quad i = 1, 2, \ldots, c$$

Heat balance:

$$H_1 F_1 + H_2 F_2 = H_3 F_3$$

Pressure balance:

$$P_3 = \min (P_1, P_2)$$

Mole fraction sums:

$$\Sigma x_{i,1} = 1, \ \Sigma x_{i,2} = 1, \ \Sigma x_{i,3} = 1$$

Physical property

$$H_1 = f_H(T_1, P_1 \, x_{1,1}, \ldots, x_{c,1})$$
$$H_2 = f_H(T_2, P_2, x_{1,2}, \ldots, x_{c,2})$$
$$H_3 = f_H(T_3, P_3, x_{1,3}, \ldots, x_{c,3})$$

We have written $c+8$ equations in the $3c+12$ unknowns $x_{i,1}$, $x_{i,2}$, $x_{i,3}$ ($i = 1, 2, \ldots, c$), F_1, F_2, F_3, P_1, P_2, P_3, T_1, T_2, T_3, H_1, H_2, H_3; the number of degrees of freedom is therefore

$$d = 3c + 12 - (c+8) = 2c + 4$$

This number corresponds to our intuition which says if the feed streams are fully specified, the outlet is fully specified. Each inlet stream has $c+2$ degrees of freedom as we proved in the previous section. It would be convenient here to write our unit model diagrammatically as in figure 5.3.

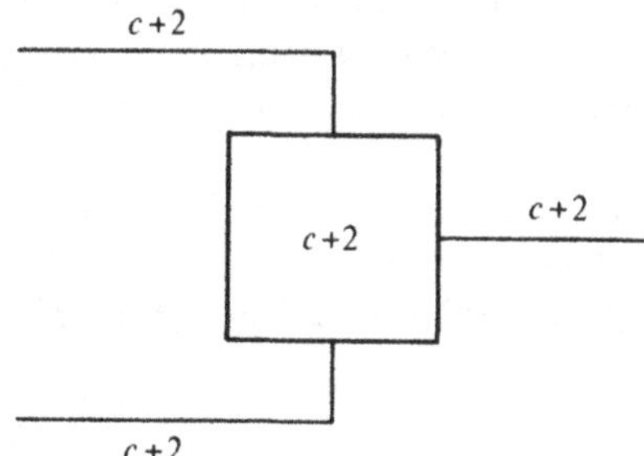

Figure 5.3. Basic block diagram model for the mixer unit in figure 5.2.

The $c+2$, as numbers on the streams, indicate the number of degrees of freedom for each stream. The number $c+2$ inside the block indicates $c+2$ relationships exist among these variables within this unit. We, however, just wrote $c+8$ relationships in $3c+12$ variables which is not consistent with this notation. The net degrees of freedom are the same:

$$3c + 12 - (c+8) = 2c + 4$$
$$3(c+2) - (c+2) = 2c + 4$$

The difference is that we defined our model using more than the minimum number of variables for each stream. In particular, we used the six variables $x_{c,1}$, $x_{c,2}$, $x_{c,3}$, H_1, H_2 and H_3 in excess of the fewest needed. We then wrote the three mole fraction sum equations and the three physical property definition equations. These variables and equations cancel each other in number, giving us the basic block diagram of figure 5.3, which indicates the 'net' number of relationships among the fewest number of stream variables needed.

One additional observation is that we are not at liberty to choose any combination of $2c+4$ stream variables to be the decision variables, that is, as the ones we specify by other means. It is obviously permissible to choose the $2c+4$ variables for the input streams as we can readily devise a means to use the model equations to calculate the $c+2$ variables for the output stream. We could also choose to specify, as the decisions, $c+2$ variables of one input stream and $c+1$ variables, excluding the pressure, of the outlet stream. The pressure balance equation as written above cannot be reversed, in general, since, if P_1 is specified, and if $P_3 = P_1$ we can say nothing about P_2 except that it is greater than, or equal to, P_1. Also we cannot specify the total flows of all three streams. These flows are related by an overall material balance (which can be derived from the equations we have written)

$$F_1 + F_2 = F_3$$

so only two of them at most can be specified. *We can choose a set of $2c+4$ stream variables to be decisions for our mixer unit only if we can then produce a solution procedure for the $c+2$ model equations to solve for the remaining $c+2$ variables. All other combinations are not legitimate choices.*

The second unit we shall consider will be the simple flash unit in figure 5.4. This unit has an obvious difference when we compare it to our mixer. We have permitted a heat input Q to exist. Rather than write

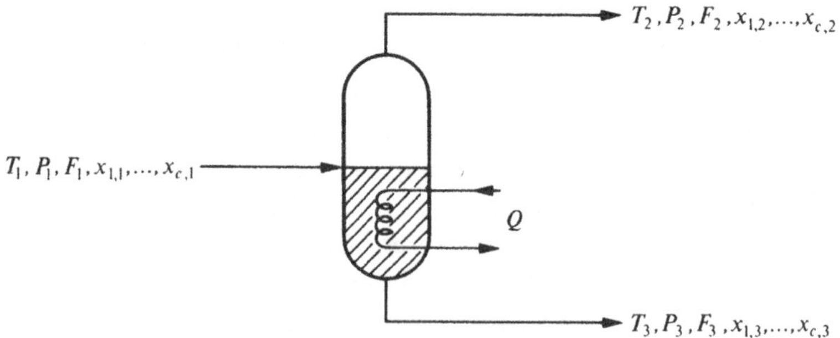

Figure 5.4. A simple flash unit.

the flash equations, which we have done before, we shall use our intuition to determine the basic block diagram model for this unit. If phase equilibrium is assumed in this unit then we know that if the heat input is specified, the outlet streams are fully specified. Thus the basic block diagram in figure 5.5 represents this flash unit. It has three lines with the numbers $c+2$ on them representing the inlet and outlet

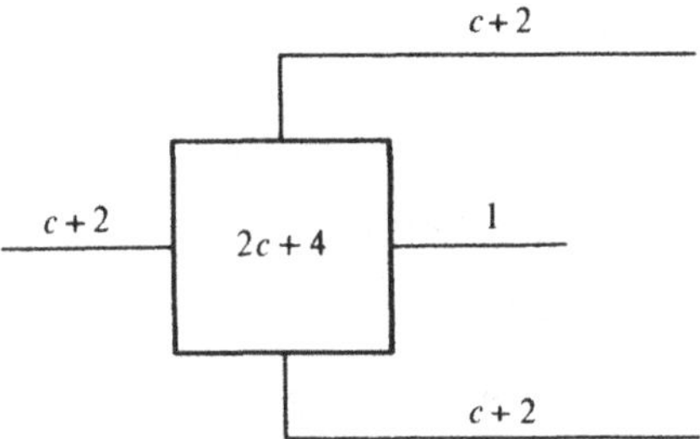

Figure 5.5. Basic block diagram model for the flash unit in figure 5.4.

streams. It has one line indicating the permitted input of heat Q with a 1 on it. Within the block is the net number of relationships, $2c+4$, that we know must exist for this unit.

Again, we cannot arbitrarily choose the decisions. For example, the temperatures T_2 and T_3 must be equal to each other and cannot be independently specified. It is possible, if we specify the temperature and pressure for the flash, that the value for Q will then be calculated and cannot be specified as a decision.

Our third unit will be the reactor unit in figure 5.6. Our analysis on the degrees of freedom for a stream in section 5.2 will greatly reduce the

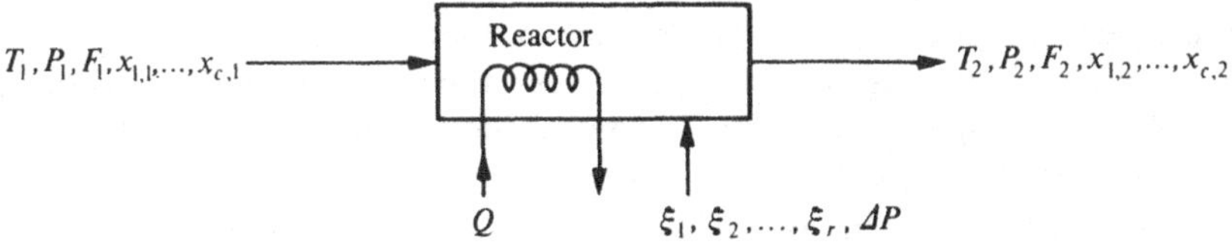

Figure 5.6. An extent-of-conversion reactor model.

effort required here to analyse a reactor unit. To analyse this unit, one will have to say in some detail the permitted design decisions to be incorporated within the model. For example, will it have a heat input (or withdrawal) allowed, can the hold-up time be varied, and so forth. Also, is it a CSTR (continuous stirred tank reactor), a tubular reactor, a combination of these, or what? We shall keep our model simple. A commonly used reactor model is called an 'extent of conversion' model. Rather than assume that the chemical reactions exit at reaction equilibrium, the user specifies (a) the r independent reactions to occur within the reactor by giving the stoichiometry for them and (b) the extent of reaction ξ_i for each reaction. From our discussion of Duhem's theorem in section 5.2, we see that we are simply trading the r equilibrium relationships that will not hold

$$\sum_{j=1}^{c} v_{ij}\mu_j = 0 \quad i = 1, 2, \ldots, r$$

for a set of user-supplied reaction extents

$$\xi_i \text{ given } \quad i = 1, 2, \ldots, r$$

If phase equilibrium is still assumed, all the same equations result with the exception of the above trade of r equations for r other equations. The sets of variables are the same. Many reactors have heating or cooling possible, so again we could allow heat to be put in or be removed. Two equations not listed when proving Duhem's theorem can still be written.

Heat balance:

$$H_2F_2 = H_1F_1 + \sum_{i=1}^{r} (-\Delta H_R)_i \xi_i + Q$$

where $(-\Delta H_R)_i$ is the heat of reaction for reaction i.
Pressure balance:

$$P_2 = P_1 - \Delta P$$

These equations introduce two new variables, Q and ΔP, and thus do not alter the degrees of freedom. Our basic block diagram model is illustrated in figure 5.7 for this reactor model.

It is not uncommon for the heats of reaction to be unavailable through the physical property package. In theory they could be; in practice they may not be. Therefore the user of this module may have to supply estimates for them; this would add r additional parameters to the unit parameter line labelled with $r+2$ in Figure 5.7.

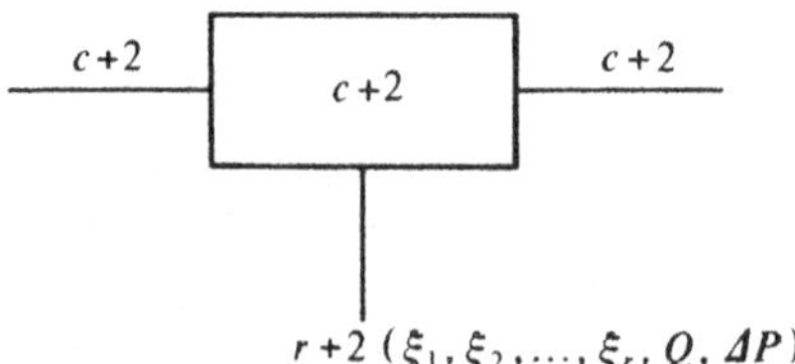

Figure 5.7. Basic block diagram model for extent-of-conversion reactor in figure 5.6.

The fourth module we shall consider is a simple stream splitter. The only task this unit can handle is to split the flow of a stream into s parts. We can draw the basic block diagram for it directly: figure 5.8. The unit parameter line is labelled $s-1$, the number of independent split

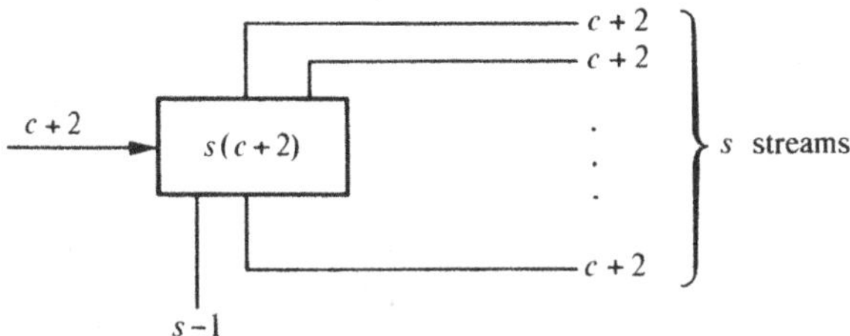

Figure 5.8. Basic block diagram for a simple stream splitter.

fractions one can specify for a splitter unit. The flows split according to the equation

$$F_j = a_j F_1 \quad j = 2, 3, \ldots, s+1$$

but the a_js are related by the equation

$$\sum_{j=2}^{s+1} a_j = 1$$

so only $s-1$ of them are independent. The $s(c+2)$ equations comprise the s overall flow balance equations just given, plus the following $s(c+1)$ equations:

$$T_j = T_1, P_j = P_1, x_{k,j} = x_{k,1}$$
$$j = 2, 3, \ldots, s+1 \quad k = 1, 2, \ldots, c-1$$

The fifth unit-type we shall consider is a counter-current heat exchanger module. Figure 5.9 illustrates this unit type along with its basic block

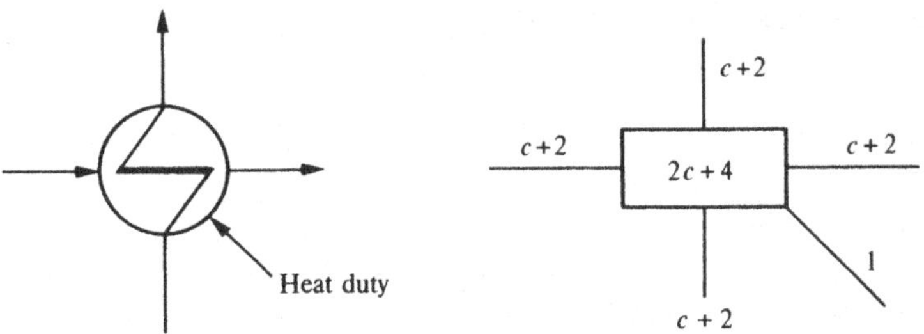

Figure 5.9. A simple heat exchanger module and associated basic block diagram model.

diagram. For this unit we permit the heat duty to be a variable which may be calculated or input.

The sixth unit is a pressure-changing device at constant enthalpy and is illustrated in figure 5.10. The pressure change will be the unit

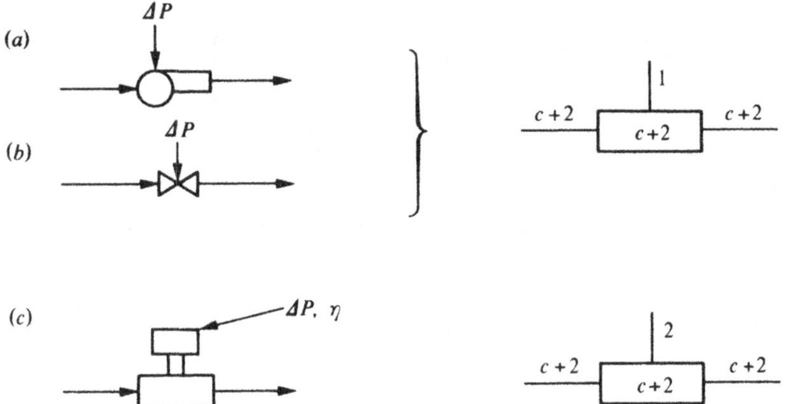

Figure 5.10. Simple pressure-changing units and their basic block diagrams: (*a*) pump, (*b*) valve, and (*c*) compressor.

parameter allowed, and this device will model a valve, and approximately model a pump. A second pressure-changing device is a compressor, also illustrated in figure 5.10. This simple compressor has the unit parameters of ΔP and efficiency permitted. The efficiency is the ratio

$$\eta = (\Delta H)_S / (\Delta H)_{\text{actual}}$$

where $(\Delta H)_S$ is the enthalpy change at constant entropy and $(\Delta H)_{\text{actual}}$ is the enthalpy change actually encountered.

The last module we shall consider is a special one. Suppose the engineer analysing a flowsheet wishes to add a specification (involving two or more variables) on a stream. An example might be

$$x_{1,i} = 3x_{2,i}$$

which says that the ratio of the mole fraction of the first component in stream i to the mole fraction of the second component must be 3. This requirement might occur when one wishes the feed to a reactor to have some components in a stoichiometric ratio. A more general specification set may comprise n equations involving i_1 variables from one stream, i_2 from another and so forth. Figure 5.11 gives a basic block diagram structure which will handle this requirement.

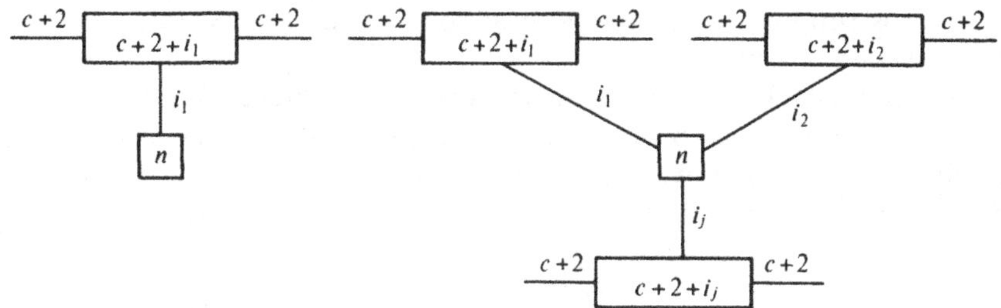

Figure 5.11. A basic block diagram module for a set of n equations connecting stream variables in several streams.

5.4 Degrees of freedom in a flowsheet

We shall consider this problem by analysing the simple flowsheet in figure 5.12. Of interest, beyond the degrees of freedom, will be to investigate how we might derive a solution procedure for this flowsheet, and this topic will be covered in chapter 7.

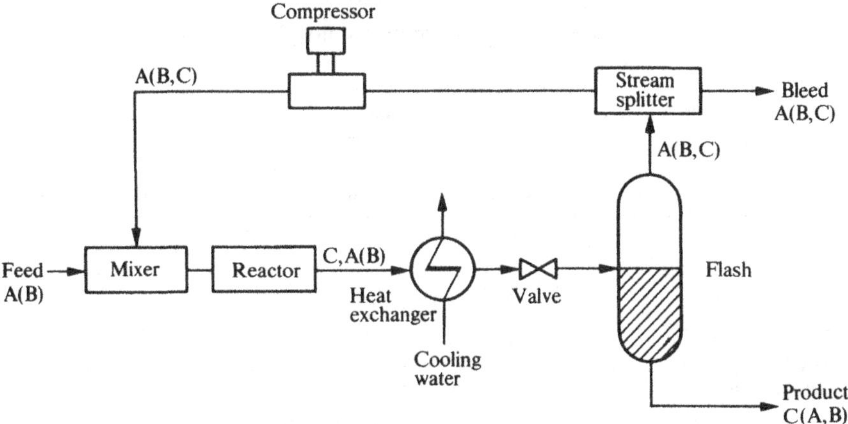

Figure 5.12. A simple flowsheet.

The flowsheet has a high-pressure feed stream of gaseous component A contaminated with a small amount of B. It mixes first with a recycle stream consisting mostly of A, and passes into a reactor where an exothermic reaction to form C from A takes place. The stream is cooled to condense component C and passed through a valve into a flash unit. Here most of the unreacted A and the contaminant B flash off, leaving a fairly pure C to be withdrawn as the liquid stream. Part of the recycle is bled off to keep the concentration of B from building up in the system.

The rest is repressurized in a compressor and mixed, as stated earlier, with the feed stream.

We can draw a basic block diagram for this flowsheet, figure 5.13, and directly ascertain the total number of degrees of freedom for it. We labelled each stream with $c+2$ in the basic block diagrams in the previous section, where c reflects the number of components in the

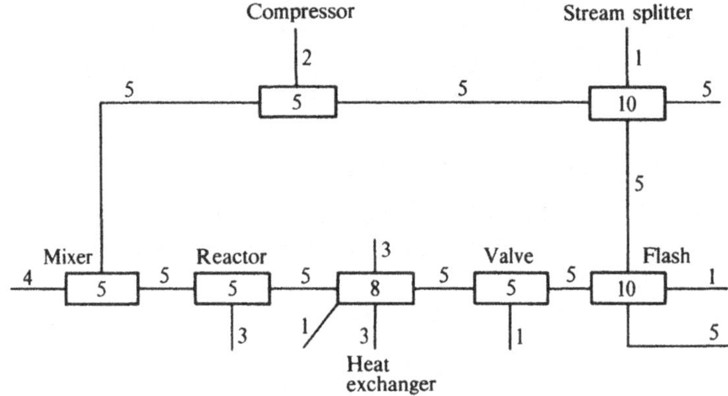

Figure 5.13. Basic block diagram for the flowsheet in figure 5.12.

Table 5.1. *Degrees-of-freedom analysis for flowsheet in figure 5.1*

Unit	No. of equations	No. of inlet stream variables	No.of unit parameters
Mixer	5	9	0
Reactor	5	5	3
Cooler	8	8	1
Valve	5	5	1
Flash	10	5	1
Bleed	10	5	1
Compressor	5	5	2

Streams leaving	No. of outlet stream variables
Cooling water	3
Product	5
Bleed stream	5

Total no. of equations	48	
Total no. of stream variables		55
Total no. of unit parameters		9
Total no. of variables		64

The flowsheet has $d = 64-48 = 16$ degrees of freedom

stream. The feed stream has two components, the cooling water stream has one, and the remaining streams have three. The numbers of equations for the mixer and the heat exchanger reflect this method of labelling. The number of equations in a mixer is equal to the number of components in the outlet stream plus two. For the heat exchanger, again the number of equations is equal to the sum of the variables in the two streams leaving, in this case, five plus three.

The results for the degrees of freedom analysis are listed in table 5.1. The flowsheet has 16 degrees of freedom and determining this number was obviously very straightforward.

6

The sequential modular approach to flowsheeting

Section 2.3.2 introduced the sequential modular approach to flowsheeting. The basic philosophy of the approach is that each unit model written for the system will calculate all output stream values given values for all input streams and for all equipment parameters. Also fundamental to the approach is that recycle stream variables will be the iterated variables at the flowsheet level of calculation. The unit modules will be solving the set of nonlinear equations which model the unit, and will in general also contain iterated variables internally.

To review some of the ideas in sections 2.3.2 and 2.3.5, we can again consider the obvious advantages and disadvantages for this approach. The major advantages result from the very clear-cut statement of how a unit module will be written. The person or team of persons writing the flowsheeting system have a very straightforward problem definition. This system is much less difficult to write than one which must handle very complex information flow. The user of the system also knows that he must supply values for all the input streams and all the equipment parameters; no other options are permitted. This limitation has the peculiar advantage of being well defined. The unit modules can be written with considerable effort put into them to make sure they are robust – that is, they can almost always succeed in calculating the output stream values. This robust quality may require considerable, at times *ad hoc*, tests to trap divergence and so forth. Also the routines can contain diagnostics to trap user errors in equipment parameter specification by checking for reasonable values.

The major disadvantage lies with the user trying to make the simulation meet his specifications and many of these specifications may be on stream values internal to the flowsheet or on exit stream values. In the sequential modular approach these must be calculated as the output from a unit module and cannot be supplied directly. We shall, in section 6.2.3, discuss the notion of a control block, the means developed by the majority of these systems to overcome this problem. Also many of these systems are developing the ability to handle more general stream types and these form a part of the approach and are again to handle user specifications which are not on the input streams and equipment parameters.

In this chapter we shall use the CONCEPT system of the CADC (1973*a*) as a particular example of a sequential modular system. It will aid us by providing a specific approach, style of input and so forth. We shall use the system to model a relatively simple flowsheeting problem. Most such systems are very similar in construction although they must clearly vary immensely in usefulness.

6.1 The solution of an example flowsheeting problem

Figure 6.1 is a flowsheeting problem we shall consider in this section. It is a particular example of the flowsheet analysed in chapter 5. The propylene feed is 2 mole % propane. In the reactor the propylene is being converted at 20 bar and 200 °C to 2-hexene by the reaction $2C_3H_6 \rightarrow C_6H_{12}$.

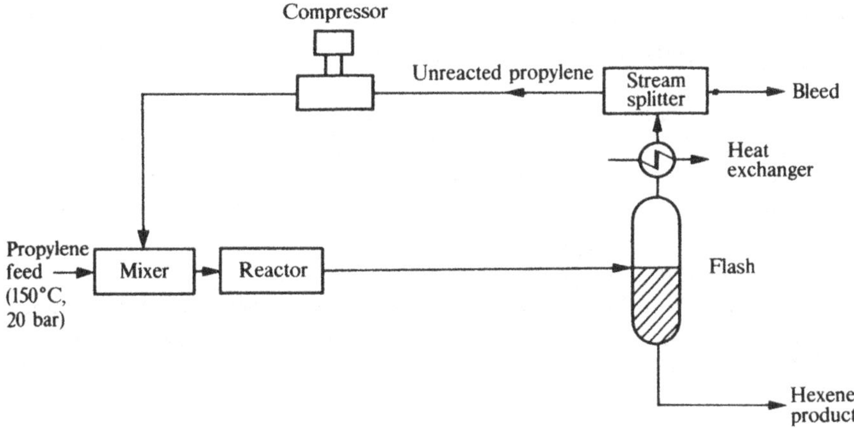

Figure 6.1. Flowsheet for example problem.

Conversion of 80% of the propylene is anticipated. The flash separator recovers unreacted propylene and much of the propane, which acts as an inert substance here. The vapour can be heated slightly to prevent condensation occurring later in the system; a drop in pressure could suffice also for this task. The bleed is to prevent too high a build-up of propane. The problem is to analyse this system. An optimization problem occurs here. The trade-off is the purity of the hexene product versus the amount of material recycled and the fraction of that recycle bled off to limit the build-up of propane. By operating the flash unit to maximize hexene purity in the liquid product stream, we must increase the flash temperature and/or reduce the pressure. In this way most of the propylene and propane leave in the vapour, but then too a sizable amount of hexene will vaporize.

A lower pressure for the flash increases the compressor duty. Also a large reflux rate will increase the cost of every unit because of the added flow each has to handle. If the hexene vaporized is a sizable fraction of the recycle, the bleed stream could carry a considerable amount of hexene product out with it.

The point of interest is to discover the possibilities for this equipment configuration. This analysis could motivate us to investigate whether a plate column would be a better investment for our separation unit. We shall use the CONCEPT system to perform the heat and material balance.

The first task is to convert the flowsheet in figure 6.1 to the form we intend to simulate. We shall call this form a functional block diagram. It must contain all the functional modules and their interconnecting streams which we are going to use to model this process. To perform this task one has to examine the list of functional models provided by the flowsheeting system that one intends to use. We find in CONCEPT the following directly useful modules among those available:

SOURCE Plant input stream source unit
OUTPUT Plant output stream destination unit
MIXER   Mixer unit for any number of streams
REACT1 Reactor with specified conversion and exit temperature –
        for a single reaction
IFLASH Isothermal flash drum
DIVIDE Simple stream splitter into any number of output streams
COMPRS Polytropic compressor with specified exit temperature
HEATER Resets a stream temperature and works out heat duty
        required to do so

With these functional modules available, we can construct the functional block diagram in figure 6.2. The CONCEPT system requires every stream in the plant simulation model to be named and to have a source and a destination module. Also every unit is named by the user as he wishes; in a separate step, the particular CONCEPT simulation module to be used to model it is designated. On our functional block diagram each stream is named, for example the reactor feed stream is called RFEED. All streams have a source and destination module; the stream PROPENE flows from FEED to MIX. On top of each module we have indicated the simulation module we intend to use. For the reactor we shall use REACT1.

Associated with the simulation module description is the number of input and output streams the module expects, the nature of the stream (e.g. the flash has two outputs and the first named will be treated as the vapour stream, the second as the liquid), and the equipment parameters

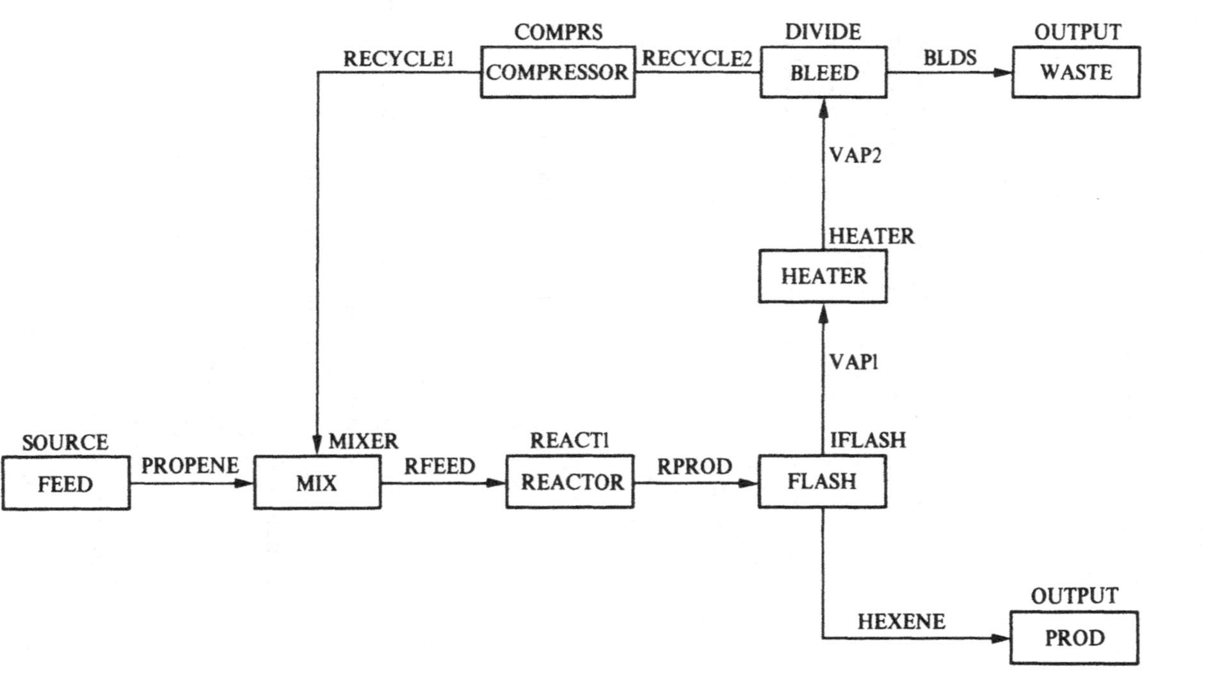

Figure 6.2. Functional block diagram for example problem.

which the user must supply for the simulation. For the modules chosen we find the following items are needed.

SOURCE   No inputs, one output, no parameters

MIXER N inputs (where N is arbitrary), one output, no parameters

REACT1   One input, one output, $c+3$ parameters

Parameters: 1 Stoichiometric coefficient of component 1
         2 Stoichiometric coefficient of component 2
         .
         .
         .
        c Stoichiometric coefficient of component c
      $c+1$ Conversion of key component
      $c+2$ Identifier of key component
      $c+3$ Outlet temperature of reactor

IFLASH   One input, two outputs (first is vapour, second is liquid), two parameters

Parameters: 1 Flash vessel temperature
         2 Flash vessel pressure

DIVIDE   One input, N outputs (where N is arbitrary), N parameters

Parameters: 1 Fraction of input stream flow to first output stream
         2 Fraction of input stream flow to second output stream
         .
         .
         .
       N Fraction of input stream flow to Nth output stream

COMPRS   One input, one output, four parameters

Parameter: 1 Output pressure
        2 Isentropic efficiency (default is 0.75)
        3 Mechanical efficiency (default is 0.94)
        4 C_p/C_V (calculated if not specified)

HEATER   One input, one output, two parameters

Parameter: 1 Required outlet temperature
        2 Pressure drop

Input of the data is in the following phases:

1. *Input of the flowsheet.* Each unit is listed together with the number of input and output streams and their names. An example would be:

REACTOR
Number of inputs: 1
Number of outputs: 1
I/P no 1: RFEED
O/P no. 1: RPROD

2. *Analysis of the flowsheet to partition flowsheet into smallest groups which have to be solved together.* CONCEPT performs this analysis for us. For our problem the order is:

    (*a*) FEED
    (*b*) MIX, REACTOR, FLASH, HEATER, BLEED, COMPRES-SOR
    (*c*) WASTE
    (*d*) PROD

3. *Analysis of each group of two or more units to determine tear streams which will be used for iteration.* For our problem, the system selected RECYCLE1 which was accepted. Thus only this one stream has to be iterated to solve the group of units listed above in (*b*).

4. *Chemical species.* The CONCEPT library contains propylene and propane. It does not contain 2-hexene but does contain 1-hexene. For this exploratory analysis, we assume we can use the properties for 1-hexene to represent those of our product. Thus we specify the components to be propane, propylene, and 1-hexene. We also tell the system to fetch the physical property parameters for these components from the library for the system.

5. *Unit simulation modules.* We specify for each unit in our functional block diagram the name of the CONCEPT simulation module that we wish to use. For MIX we specify MIXER, and so forth.

6. *Unit equipment parameters.* For each unit we must supply the actual values for the parameters needed. The following is an example set.

    REACTOR: 0., 2., −1., 0.8, 51, 200.

    These tell the unit the reaction is

    0 propane + 2 propylene → 1 hexene

    Conversion is 0.8 of the entering propylene (component 51 in the CONCEPT library of components) and the temperature is 200 °C for the outlet stream.

    FLASH: 130., 12.

    The flash temperature is 130 °C and pressure 12.0 bar.

HEATER: 130., 0.

The heater outlet temperature is 130 °C (no change so the unit is being ignored here) and pressure drop is 0 bar.

BLEED: 0.02, 0.98

At this point, 2% of the flash vapour will be bled off.

COMPRESSOR: 20., 0.75, 0.94, 1.26

The outlet pressure is 20 bar, efficiencies are 0.75 and 0.94, and C_P/C_V is 1.26 for the stream.

7. *Convergence tolerance and promotion.* We specify that convergence should be assumed if the largest difference in the torn stream variables guessed and calculated is 0.2%, that in any case a maximum of ten iterations should be done, and that the system should use the Wegstein (1958) convergence-acceleration method to reguess the new tear stream values from the old ones.

8. *Input stream values.* We must give the input stream flowrate (1000 mol/h), temperature (150 °C), pressure (20 bar), and composition 2% propane, 98% propylene and 0% hexene).

9. *Tear stream guesses.* We must specify guesses for the stream RECYCLE1: flowrate (200 mol/h), temperature (150 °C), pressure (20 bar) and composition (25% propane, 75% propylene, 0% hexene).

With this input, which is typical of the input needed for any sequential modular flowsheeting system, the problem can now be simulated. When simulated it reached a convergence tolerance of 0.65% in ten iterations which, for our purposes, is sufficiently accurate and we accept the answers. Since CONCEPT saves its results in a data file as well as prints them, we could equally well have requested another ten iterations from the current results to get a tighter tolerance.

The output from the system repeats the input data for verification, gives a history of the calculations as they proceed, and presents the final converged values for all the streams (see figure 6.3) and units. Some of the units provide useful results not needed to calculate the stream values, e.g. the heater writes out the heat duty which results for it. The output will also contain diagnostic messages if appropriate. Again this type of output is typical of most of these systems.

The system is most useful at this point because once the problem is set up and running, we can vary a number of the equipment or stream data values to see their effect. For this system the flash temperature was run at values of 105 °C, 120 °C, 130 °C and 140 °C. Also at 105 °C, the flash

	RECYCLE	PROPENE	RFEED	RPROD
FLOWRATE	+4.0572E2	+1.0000E3	+1.4047E3	+9.4420E2
TEMPERATURE DEGC	+1.8976E2	+1.5000E2	+1.6362E2	+2.0000E2
PRESSURE BAR	+2.0000E1	+2.0000E1	+2.0000E1	+2.0000E1
VAPOUR FRACTION	+1.0000	+1.0000	+1.0000	+1.0000
ENTHALPY KJ/KGM	+3.6513E4	+2.1701E4	+2.5967E4	+4.1678E4
—MOLE FRACTIONS—				
PROPANE	+1.2913E−1	+2.0000E−2	+5.1197E−2	+7.6165E−2
PROPYLENE	+4.2213E−1	+9.8000E−1	+8.1952E−1	+2.4384E−1
1−HEXENE	+4.4875E−1	+0	+1.2928E−1	+6.8000E−1

	VAP2	RECYCLE2	BLDS	HEXENE
FLOWRATE	+4.1364E2	+4.0537E2	+8.2728	+5.3057E2
TEMPERATURE DEGC	+1.3000E2	+1.3000E2	+1.3000E2	+1.3000E2
PRESSURE BAR	+1.2000E1	+1.2000E1	+1.2000E1	+1.2000E1
VAPOUR FRACTION	+1.0000	+1.0000	+1.0000	+0
ENTHALPY KJ/KGM	+2.9086E4	+2.9086E4	+2.9086E4	+1.6218E4
—MOLE FRACTIONS—				
PROPANE	+1.2885E−1	+1.2885E−1	+1.2885E−1	+3.5092E−2
PROPYLENE	+4.2240E−1	+4.2240E−1	+4.2240E−1	+1.0462E−1
1−HEXENE	+4.4875E−1	+4.4875E−1	+4.4875E−1	+8.6029E−1

	VAP1
FLOWRATE	+4.1364E2
TEMPERATURE DEGC	+1.3000E2
PRESSURE BAR	+1.2000E1
VAPOUR FRACTION	+1.0000
ENTHALPY KJ/KGM	+2.9086E4
—MOLE FRACTIONS—	
PROPANE	+1.2885E−1
PROPYLENE	+4.2240E−1
1−HEXENE	+4.4875E−1

Figure 6.3. Final stream data for example problem.

pressure was dropped to 10 bar, from 12 bar, and run. The effects on recycle flowrate and hexene purity are quite substantial, as table 6.1 shows.

If we were to do a costing and economic analysis, we could then compare the costs and projected value for these alternatives. For costing purposes, we would need to provide sufficient data so that the units could be sized, the construction materials selected and so forth. The flash unit might be sized to give five minutes hold-up for the liquid portion with an equal volume for the vapour disengagement section. This hold-up affects dynamic behaviour of the whole process, level-controller action for the unit, process reliability and so forth and has to

Table 6.1. *Results of five different runs for example problem*

Run	1	2	3	4	5
Flash T (°C)	105	105	120	130	140
Flash P (bar)	10	12	12	12	12
Recycle flowrate (mol/h)	179	120	264	406	630
Hexene product purity	81.2%	41.6%	82.0%	86.0%	89.5%

be determined by other considerations. Once decided the unit can be sized. The temperature and pressure can aid to select materials (very likely carbon steel here) and wall thicknesses.

The heat balance for the flash will indicate the cooling needed for it, or its feed, so a cost for utilities can be obtained. Similar results are available from the other units. This costing is a postprocessing step; that is, it is done after the flowsheet heat and material balances are satisfied. It may be done within the run by a set of costing modules or done after the run but using the data file just created. The CONCEPT system, for example, produces a data file which can be accessed by a separate system called ECONOMIST. With this system one can cost the units, assume various product and raw material prices, and carry out a number of different economic analyses.

6.2 Other features

In the last section we solved a simple flowsheet, going into some detail on how it would be set up for the flowsheeting system and indicating how one might then use the system to investigate the flowsheet. We used the system in its most conventional form, where all the chemical components and unit modules were in the library. Also, all user-supplied specifications were for the input stream and unit equipment parameter values. We shall discuss in this section how the typical sequential modular systems are, or might be, designed to handle less conventional cases.

6.2.1 User-supplied chemical component data

This topic has already been discussed in chapter 4. Basically, the parameters used by the physical property routines to calculate mixture properties have to be supplied or estimated.

6.2.2 User-supplied unit modules

The computer environment in which a unit module will find itself has to be clearly defined for the user who wishes to write his own unit module. He must know exactly how the executive system will 'call' the routine, how it will supply the input stream values, where he is to put the output stream values, etc. He, of course, must be able to program in the basic language for the system, usually FORTRAN. Table 6.2 gives some of the data that the CONCEPT executives supply to a user. It indicates the type of data one might expect to see in most systems of this type.

In addition the unit equipment parameters have to be accessible for reading from and writing to (if used for intermediate storage). Figure 6.4 illustrates these ideas.

Table 6.2 *CONCEPT data passed to and from a unit module*

System data

NC	No. of components
NV	No. of variables stored for a stream
MODE	If 0, initial entry to unit, so output stream values have never been calculated before and therefore do not exist for use If 1, the unit must be in an iteration loop and its previous output stream values are available If 2, all recycle variables have converged and any special final output should be produced (e.g. for the HEATER unit in the last section, the heat duty should be calculated and printed)
NIN	The number of input streams to the unit
NOUT	The number of output streams from the unit
ICH	All final results should be written onto file ICH
ICH1	All intermediate messages should be written to file ICH1. This output aids in subsequent program checkout

Stream Data

STMI	A vector of data containing the input stream values
STMO	A vector of data for containing the output stream values

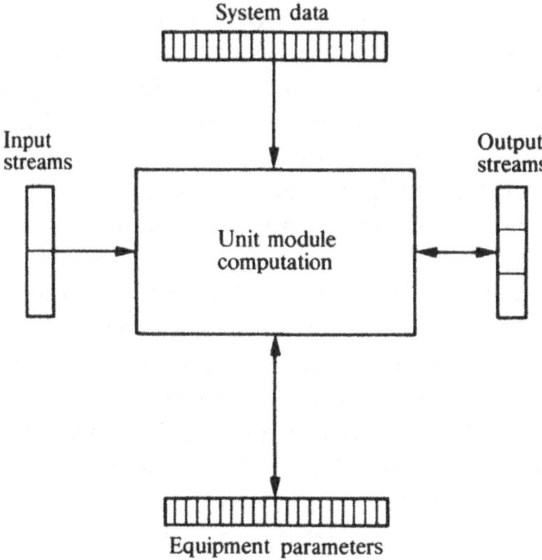

Figure 6.4. The unit module environment.

Errors from which no recovery is possible may occur within the routine, and a means to exit and stop the flowsheeting program is also needed.

The last type of information needed is how to access the physical properties package for physical property data. We have already discussed the kind of physical property support that the user might expect in chapter 4. The typical access is to pass the stream data – usually T, P and composition – to the physical property package, together with an indicator as to which property or set of properties one wishes supplied in response to the request.

Armed with this type of information, the user can produce a user-written module to do his special calculations. Reactors commonly fall into this category since they tend to be very special purpose. For example, they may require the integration of a set of rate equations over the length of a tubular reactor, where the user might want the length to be supplied as an equipment parameter so that he can easily perform parametric studies for it.

6.2.3 Control blocks and nonstandard streams

Control blocks have been established as a means to permit the user to put specifications on intermediate and/or output stream variable values. For example, the engineer may wish to have a certain product purity in the top stream of a distillation column. He could accomplish this if he could adjust an appropriate equipment parameter for the column to a value to meet this specification. The column module may have top product rate to feed rate ($\beta = D/F$) as an equipment parameter and/or it may have reflux ratio. (The former is probably a better equipment parameter in most cases since it is much more readily estimated for a column.) If the user could measure the top stream purity and adjust the chosen column equipment parameter automatically, he could meet his specification. We shall outline an approach which can be taken to meet this requirement.

The first useful notion is that of a nonstandard stream which is simply a means to transfer information from one unit to another where the information is not necessarily stream-related data. Figure 6.5 illustrates a means to accomplish this transfer where the unit module is unaware of its occurrence. This figure implies that the information input streams are handled by a preprocessing unit (which may be user written or be part of the library of modules). This unit receives the information stream data, perhaps performs some calculations based on it, and distributes the results into the equipment parameters for the unit module.

The unit module then performs its calculations based on the current input stream and equipment parameter values. It is written exactly as before.

The last step is a postprocessing unit which collects values from the equipment parameters and/or the input and output stream data. It may then perform a calculation on them, and finally it puts the results into

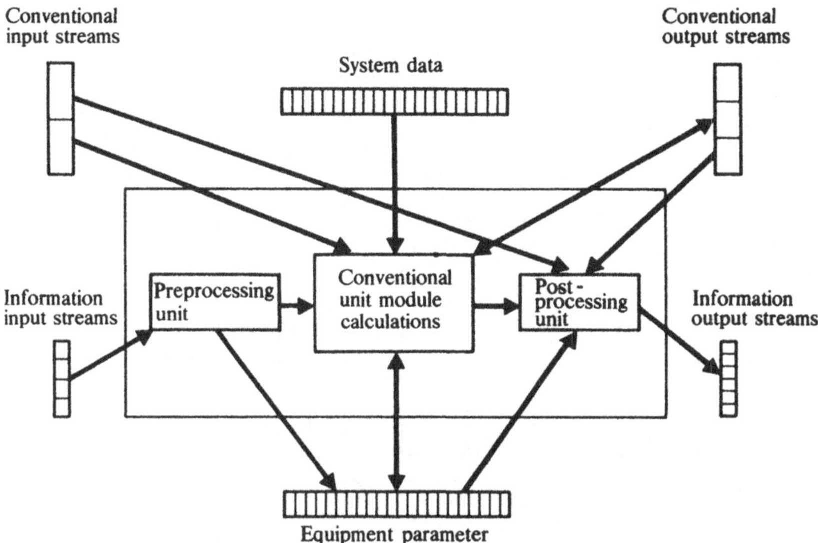

Figure 6.5. Handling information streams.

the information output streams. Again the postprocessing can be a standard one in the system or user supplied.

To use the type of structure we can return to our distillation column where a required product purity is desired for the top stream. We shall assume $\beta = D/F$, the top product to feed flowrate ratio, is in the equipment parameters for the column. Figure 6.6 illustrates the flowsheet resulting. The information stream for β creates a recycle in the flow diagram. It can be treated in identically the same manner as a recycle created by conventional streams. This recycle indicates that the value for β has to be guessed and iterated.

The reguessing of β is a duty, as illustrated here, of the controller block which performs the following calculations:

1. Compare input signal (y_3) to setpoint (SP)
    (*a*) If not equal, adjust output signal (β)
    (*b*) Otherwise leave β unchanged
2. Exit

One could use a classical three-term controller which adjusts the output by the formula

$$\beta_n = K_P E_n + K_I \sum_{i=1}^{n} E_i + K_D(E_n - E_{n-1})$$

where $E_n = SP - y_3$ at the current iteration (iteration n), β_n is the

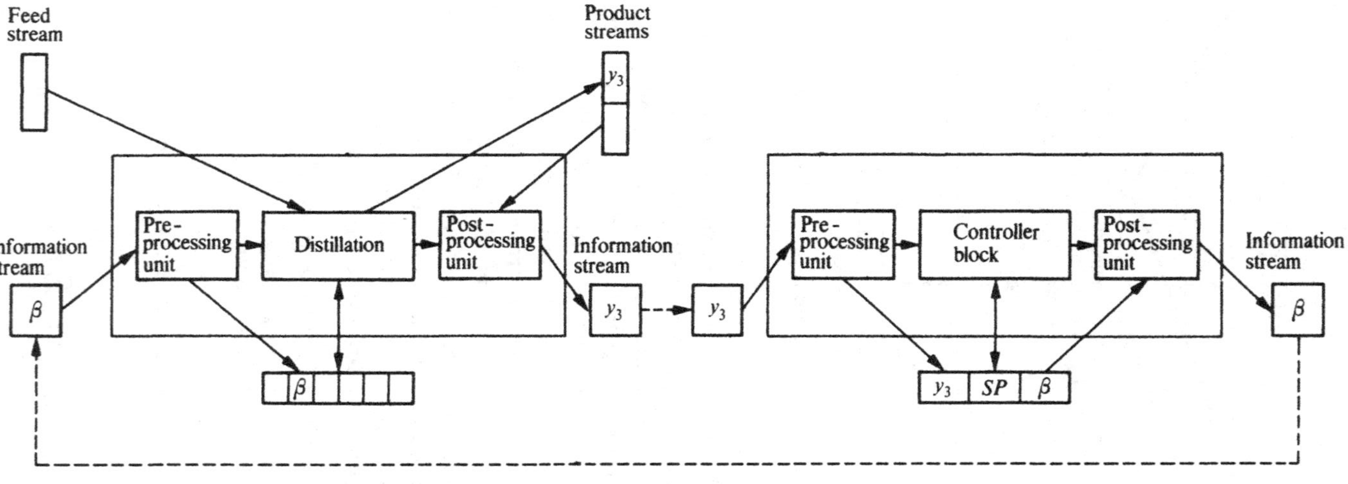

Figure 6.6. A control block for meeting a specification (SP) on top product composition for a column.

predicted output, and K_P, K_I, K_D are the proportional, integral and derivative control action constants, respectively.

The controller could also use the secant method internally to adjust β, since it is effectively trying to solve the implicit equation

$$E = SP - y_3 = f(\beta) = 0$$

We can see this by considering the following calculation.

1. Guess β
2. Calculate the top product composition for the third component, y_3
3. Calculate $E = SP - y_3$
4. Compare E to zero
    (*a*) If not essentially zero, adjust β and iterate from step 2
    (*b*) If essentially zero, continue
5. Exit

We shall consider methods to converge tear streams for a flowsheet in section 6.3. At this point we shall only warn that this latter approach (i.e. the controller uses a secant algorithm) will lead to computational difficulties if β is not the only tear variable, such as when the column itself is inside a group of units which have to be solved as a group. The method to converge torn streams may be attempting to converge them using a type of secant method. Thus the controller might adjust β using a secant method and then the system might impose a second adjustment based on a secant approximation, which computationally may be very unstable.

6.3 Convergence of tear streams

In section 3.3 we discussed in some detail the typical methods used to solve sets of nonlinear equations. We distinguished between implicit and explicit iteration schemes. For the former the tear variables are guessed and the resulting calculations give rise to error functions which must be driven to zero. In the explicit iteration case, the computations give rise to calculated values for the tear variables themselves. We can thus exploit the use of this information as the calculated values can be used directly as the next guess.

In a conventional sequential modular flowsheeting system, all stream tears are explicit. Suppose, however, we permit implicit tears to occur in addition to the explicit stream tears. The extra freedom for converging a flowsheet permitted by this option is very interesting.

The mechanism to create implicit tears is indicated in figure 6.7. Here two output values from the unit which should be zero are tied to two input values which can be adjusted to drive them to zero. The executive recognizes these tears as implicit and treats them as such. They *must* be tears. Care has to be taken that a cause and effect occurs; that is, the variables to adjust must have an effect on the tear functions. If not, it is obvious the iteration must fail.

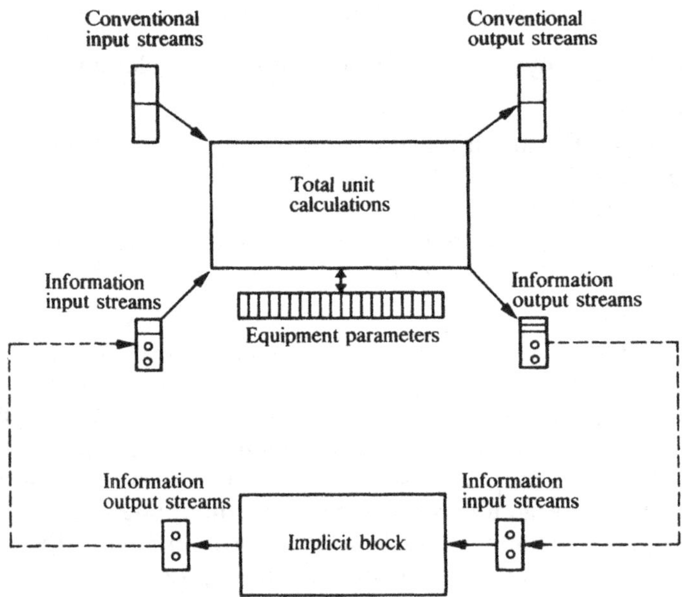

Figure 6.7. Implicit blocks in a flowsheet system. The dashed lines indicate transfer of stream data by the system executive.

With implicit tears we can now do some interesting calculations. Consider the flash unit calculation discussed in section 3.1.1. We could calculate the error $\Sigma y_i - \Sigma x_i$ and put this into an information output stream *without iterating within the flash unit to drive it to zero* (Johns, 1970). The value of $\varphi = V/F$ is in an input information stream and is tied to the error by an implicit block. The executive system will discover this implicit tear and, external to the flash routine, will adjust φ to drive the error to zero.

We might wonder what the advantages to this approach are. Figure 6.8 illustrates one. The tears for the flowsheet are the vapour recycle and the implicit loop. They can be converged *simultaneously* rather than in an imbedded fashion. The normal calculation would be

1.   Guess recycle vapour stream
2.   Calculate mixer
3.   Calculate reactor

4. Calculate flash (converge $\Sigma y_i - \Sigma x_i$ to zero by adjusting $\varphi = V/F$)
5. Compare calculated recycle stream values to those guessed
    (*a*) If not essentially equal, iterate from step 2
    (*b*) Otherwise continue
6. Exit

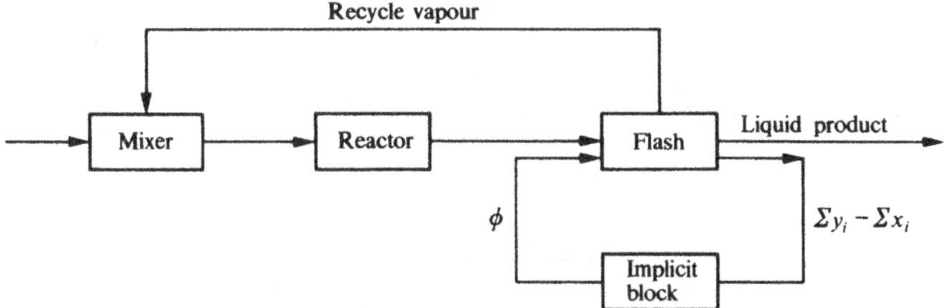

Figure 6.8. Illustrating an advantage of an implicit block.

With the implicit block we could proceed as follows.

1. Guess vapour recycle stream values
    Guess φ for flash unit
2. Calculate mixer
3. Calculate reactor
4. Perform flash unit calculations *once* through to get new values for recycle stream and error $\Sigma y_i - \Sigma x_i$
5. Compare calculated recycle stream values to those guessed and simultaneously check if error $\Sigma y_i - \Sigma x_i$ is zero
    (*a*) If convergence not achieved, iterate from step 2
    (*b*) Otherwise continue
6. Exit

In this approach the flash iteration is converged with the flowsheet recycle calculations.

We might reconsider the control block problem in figure 6.6 again. If the postprocessor for the distillation unit had the *SP* value available, it could output an error term and the user could tie it to the input information stream, supplying β through an implicit block. Figure 6.9 illustrates this.

We might now illustrate the use of information streams and implicit tears on a larger problem. Consider the flowsheet problem stated in section 5.4.

All specifications listed for intermediate and output streams will have to be satisfied by adjusting appropriate input stream and equipment

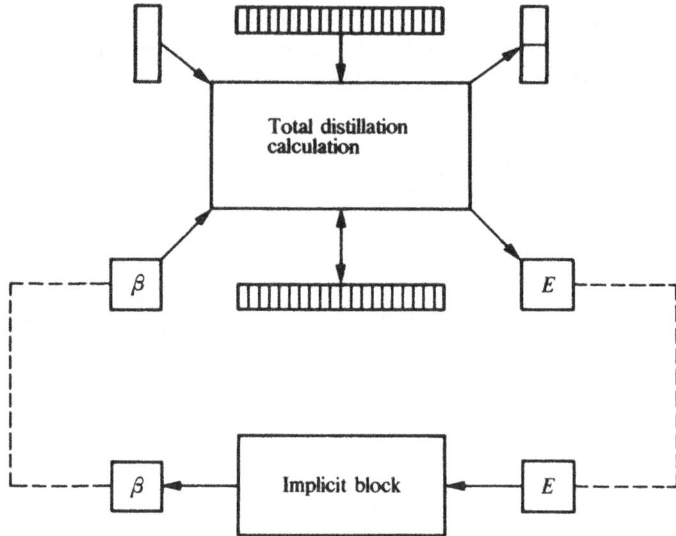

Figure 6.9. Using an implicit block as a controller block.

parameter values. The following give those specifications in this category and a set of variables we could adjust.

Intermediate stream specifications	Adjustable variables
Inlet T to reactor	Feed temperature
Outlet cooling water T	Cooling water flowrate
Product purity from flash	Conversion in reactor
Liquid stream flow from flash	Feed flowrate
Fraction recovery of C in flash	Pressure drop in valve
Outlet pressure for compressor	Pressure increase in compressor

Two of these specifications are perhaps best met by rewriting the unit modules involved. The cooling water exchanger can be rewritten to calculate the needed water flowrate to get a specified exit temperature for the water. Similarly the compressor module can be written to set the outlet pressure (as COMPRS does already in the CONCEPT system). This reduces the number of implicit tears to four and figure 6.10 gives a functional block diagram for this case. In addition the five variables for the recycle stream are a set of explicit tears, bringing the total, at the flowsheet level, to nine. We might note that in section 7.3 we shall partition this problem into four smaller groups with one tear for the first, zero each for the second and third, and three for the fourth group, for a total of only four tears.

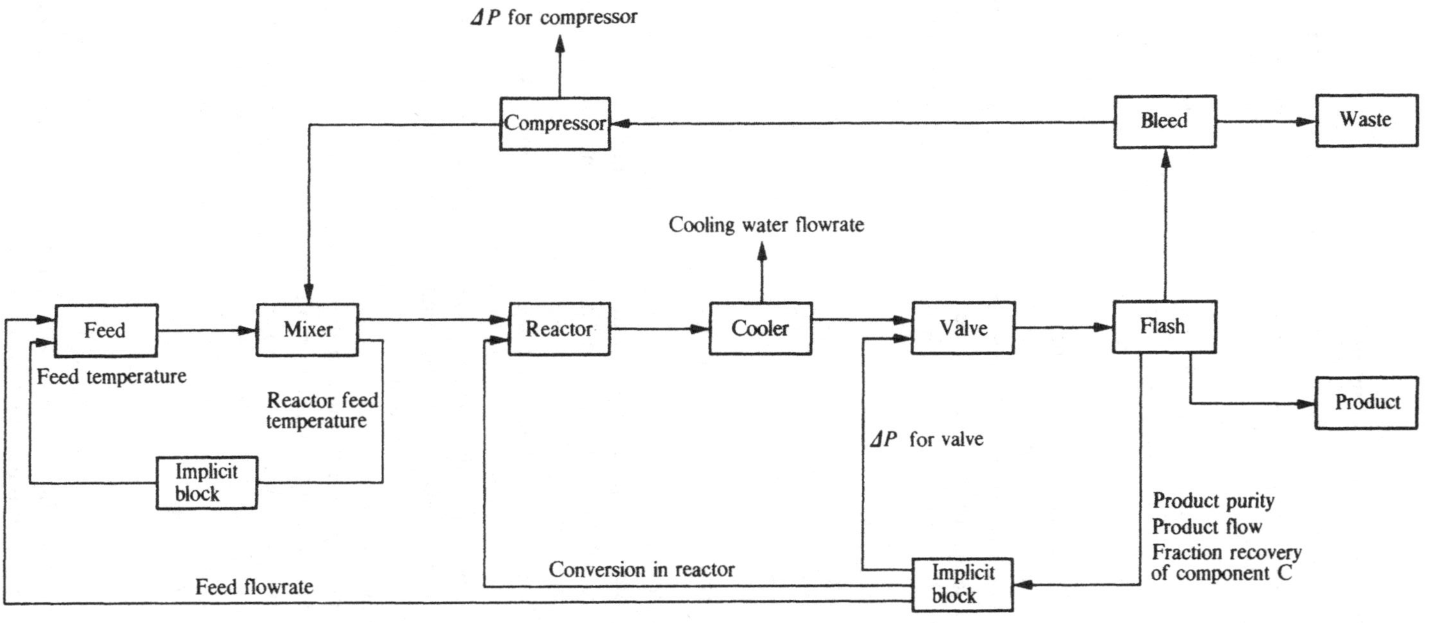

Figure 6.10. Functional block diagram for example in figures 5.12 and 5.13.

6.4 Partitioning and tearing a flowsheet

The literature contains a number of articles related to finding tear streams for a flowsheet. (A sample includes Sargent and Westerberg, 1964; Lee and Rudd, 1966; Christensen and Rudd, 1969; Forder and Hutchison, 1969; Johns, 1970; Barkley and Motard, 1972: Upadhye and Grens, 1972; Pho and Lapidus, 1973; Cheung and Kuh, 1974; Guardabassi, 1974; Janicke and Biess, 1974; Smith and Walford, 1975.) From our discussion in the previous section, we know that all implicit tears are set up by the user when he includes an implicit block in his functional block diagram. That these must be tears is known *a priori* and what remains is to locate all the other tears.

Consider the functional block diagram in figure 6.11. Assume the dotted stream is an implicit tear. The numbers associated with each stream represent the number of significant variables, i.e. ones which are not essentially zero, for that stream. We wish to determine the best (in some sense) computational sequence for this flowsheet. The task can be broken into two steps: partitioning with precedence ordering and choosing tear streams.

6.4.1 Partitioning and precedence ordering

Partitioning a flowsheet is to locate for it the groups of units which must be solved together, and precedence ordering is placing these groups in a proper sequence for computation. These groups should each contain the fewest number of units possible. For the example in figure 6.11, we can see that unit H can be calculated first and by itself. It forms group 1. The units A, B, C, D and E have a recycle structure which forces them to be calculated together, and they form group 2. Units F and G are in a recycle loop, and they form group 3. Finally unit I can be calculated and forms group 4. (A discussion similar to this one for equation sets was given in section 3.4.1.)

A quite simple algorithm which traces unit outputs (or inputs) exists (Sargent and Westerberg, 1964) and allows one to find these groups. Basically one traces from one unit to the next through the unit output streams forming a 'string' of units. This tracing continues until (*a*) a unit in the string reappears or (*b*) a unit with no more outputs is encountered. For case (*a*) all the units between the repeated unit together with the repeated unit become a group. This group is collapsed together and treated as a single unit and the tracing continues from it. For case (*b*) the top unit or group of units is placed at the top of a list of groups and is deleted entirely from the problem. This list will contain our groups in a correct order for computation when we finish.

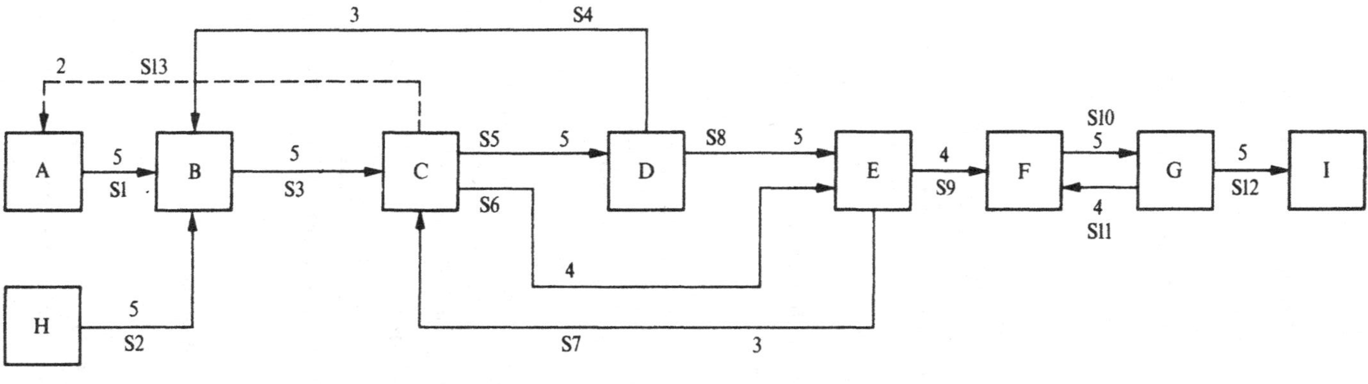

Figure 6.11. A complex functional block diagram. The dashed line indicates that S13 is an information stream which must be torn.

For our example this algorithm gives the following results:

Start with any unit first (say C).

$$\text{String 1: } C \to D \to E \to C \qquad \text{case } (a)$$

Collapse the units into a group and treat as a unit. Continue forming an output string.

$$\text{String 1: } (CDE) \mapsto B \to C \qquad \text{case } (a)$$

Unit C is already in the string. Collapse the units into a group and continue.

String 1: $(CDEB) \mapsto A \to B$ case (a)
String 1: $(CDEBA) \mapsto F \to G \to F$ case (a)
String 1: $(CDEBA) \mapsto (FG) \to I$ case (b)

The unit I has no further outputs. Put onto list of groups and delete it completely from the problem. Continue with forming string 1.

$$\text{String 1: } (CDEBA) \mapsto (FG) \qquad \text{case } (b)$$

The group FG has no further outputs. Put onto top of list of groups. Delete from problem completely.

$$\text{String 1: } (CDEBA) \qquad \text{case } (b)$$

The group (CDEBA) has no other outputs; put at top of list of groups and delete from problem completely. String 1 is now empty. Unit H has yet to be considered so we start a string with it.

$$\text{String 2: } H \qquad \text{case } (b)$$

Unit H has no feeds to any unit which still exists in the flowsheet. Delete it and put at the top of the list of groups. The flowsheet is completely deleted so we stop. Our list of groups is

$$\text{List of groups: H, (CDEBA), (FG), I}$$

which agrees with our earlier observation. This algorithm is readily programmed so partitioning and precedence ordering is very straightforward and very fast.

6.4.2 Tearing a group of units

Next we have to determine just how to solve the groups with more than a single unit in them by choosing the set of tear streams for each. For example, the group (FG) can be solved by guessing stream S10, calculating unit G then F and getting an updated set of values for S10. It can also be solved by guessing S11, calculating unit F then G and getting an updated set of values for S10. If we intend to use successive substitution the two sequences are similar:

$$\text{Sequence 1} \quad S10 \to G \xrightarrow{S11} F \xrightarrow{S10} G \xrightarrow{S11} F \ldots$$

$$\text{Sequence 2} \quad S11 \to F \xrightarrow{S10} G \xrightarrow{S11} F \xrightarrow{S10} G \ldots$$

The newly calculated value for the guessed stream becomes the input to the next iteration. Clearly the only difference is which stream is guessed to start the calculation.

If we intend to accelerate the convergence, then the sequences are quite different.

$$\text{Sequence 1:} \quad S10 \to G \xrightarrow{S11} F \xrightarrow{S10} accelerate \xrightarrow{S10} G \xrightarrow{S11} F \ldots$$

$$\text{Sequence 2:} \quad S11 \to F \xrightarrow{S10} F \xrightarrow{S11} accelerate \xrightarrow{S11} F \xrightarrow{S10} G \ldots$$

We might choose between them by picking the sequence which guesses and accelerates the fewest variables. In figure 6.11 stream S10 has five variables and stream S11 has four variables so we would choose sequence 2.

The group (CDEBA) offer a much larger challenge. We know the implicit tear S13 must become an actual tear. How many additional tears to choose is not at all obvious. For example we could choose to tear S3 and S7, using the computational order in figure 6.12: tear S13, A, S1, tear S3, tear S7, C, <u>S13</u>, S5, S6, D, S4, S8, E, <u>S7,</u> B, <u>S3</u> where the *calculated* stream values for S13, S7 and S3 are underlined. We note here that the sequence (tear S13, A) could have been moved to just before unit B, in which case the calculated value for <u>S13</u> appears before

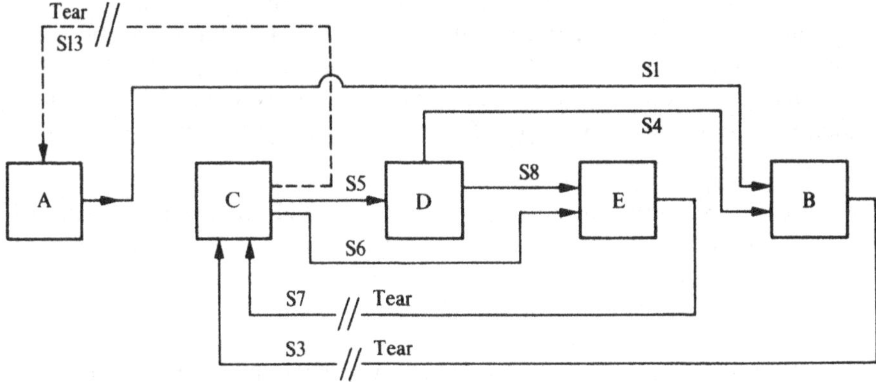

Figure 6.12. Solving group (CDEBA) using three tears: S13, S7, S3. All other streams feed forward in the sequence.

the tear 13 in the sequence. But because this tear is implicit, values obtained for S13 in C are only indirectly producing values for the input to A. They are obtained only after the executive looks at the S13 values from C to see if they are zero, and then it reguesses the inputs to A to drive these outputs from C to zero if they are not zero. Thus S13 is truly torn irrespective of the sequence used here.

Is this the best order to use? By what criteria can we judge the orders possible so that we can choose? Can an effective algorithm be implemented to select the correct order? These questions have been approached in the literature, and algorithms exist for the following criteria:

1. Choose the order giving the fewest number of tear streams. (The best algorithm is probably that of Barkley and Motard (1972)).
2. Assign a weight to each stream. This weight can reflect the expected difficulty associated with tearing the stream and may, for example, be equal to the number of significant torn variable values in each stream. Find the order which minimizes the sum of the weights associated with the torn streams. Two algorithms to handle this problem are available (Christensen and Rudd, 1969; and Pho and Lapidus, 1973).
3. Tear to give the best convergence characteristics for successive substitution. The algorithm is that of Upadhye and Grens (1975), and we shall consider it in more detail here.

Upadhye and Grens define decompositions of a flowsheet, that is the choosing of the tear streams, as belonging to two classes: redundant and nonredundant. Their work argues qualitatively and demonstrates that nonredundant decompositions have better convergence properties than redundant ones for the method of successive substitution. Fortunately it is relatively easy to locate a nonredundant decomposition for small problems using a replacement rule they define. For larger problems it could become tedious, but they describe a more appropriate algorithm in their paper for such problems. We shall apply the rule, since it is relatively easy, to the group (CDEBA). The algorithm is as follows:

1. Find any tear set for the group
2. Apply the replacement rule which states: If all the output streams of a unit are represented in the tear set, a tear set in the same family of tear sets will be produced if these output streams are replaced by the input streams for that unit
3. If a stream appears more than once in the tear set at any step, delete all but one occurrence for it. The family with the repeated stream is a redundant one. The new set is in a new family. Start over with it as the tear set

4.   Repeat steps 2 and 3 until a family is reached where no member has a repeated stream in its tear set

Figure 6.13 illustrates an application of the algorithm for our example problem. We started with our observed tear set in figure 6.12: S3, S7 and S13. Stream S3, being the sole output of unit B, can be replaced by the input streams to B, S1 and S4. Also stream S7 can be replaced by S6 and S8. The next replacement along the left branch of our 'tree' is to substitute S1 by S13 but this gives two occurrences to S13 in the tear set. We strike one out, leaving us with the tear set S4, S7 and S13. The rest of figure 6.13 is to generate all the tear sets within the family containing S4, S7 and S13. The complete tear sets which are shown with an asterisk have appeared elsewhere and require no further analysis. None gives rise to a repeated occurrence of a stream, and this family is non-redundant. If successive substitution were used, any member of this family would have the same convergence characteristics as any other member so we can now apply a secondary criterion to select among the members.

We want S13 to be torn so we can look among those sets containing S13. These are (S4, S7, S13), (S4, S6, S8, S13) and (S5, S6, S13). We could choose among these three alternatives by choosing the one which minimizes the total of the associated weights given in figure 6.11:

$$
\begin{array}{llcc}
\text{S4, S7, S13} & 3+3+2 & = & 8 \\
\text{S4, S6, S8, S13} & 3+4+5+2 & = & 14 \\
\text{S5, S6, S13} & 5+4+2 & = & 11
\end{array}
$$

This criterion gives S4, S7 and S13 as our best tear set. The order to calculate the units is found by deleting these streams from the group and applying the partitioning algorithm to the result. Since these are precisely the streams with a backward direction in figure 6.13, we can see the computational order is (ABCDE).

Motard and Westerberg (1978) present an interesting perspective on redundancy in tearing. We can define, for an irreducible group of units, the computational loops which are included within it. First we define a 'full string' which comprises a sequence of units and streams

$$(U(i_1), S(i_2), U(i_3), \ldots, U(i_{j-1}), S(i_j), U(i_{j+1}), \ldots, S(i_n))$$

such that stream $S(i_j)$ is an output stream from unit $U(i_{j-1})$ and an input stream to unit $U(i_{j+1})$. An example of a 'full string' for the example in figure 6.12 is

$$(A, S1, B, S3, C, S5) \qquad \text{(full string example)}$$

We define a unit loop for a full string as one where the first unit in the full string is about to be repeated. An example for figure 6.11 is

$$(A, S1, B, S3, C, S13) \qquad \text{(unit loop for full string)}$$

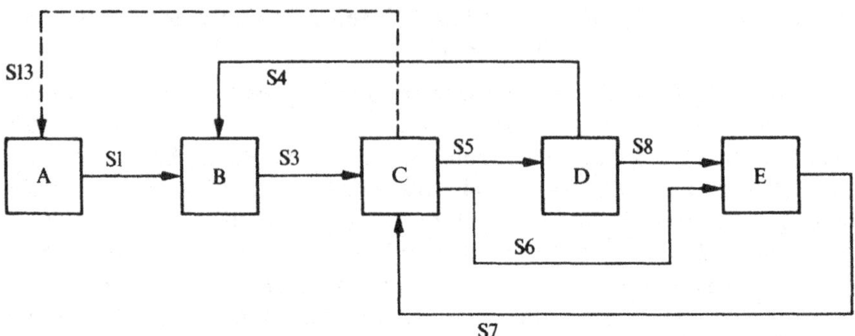

Unit	Outputs	Inputs
A	S1	S13
B	S3	S1,S4
C	S5, S6, S13	S3,S7
D	S4, S8	S5
E	S7	S6,S8

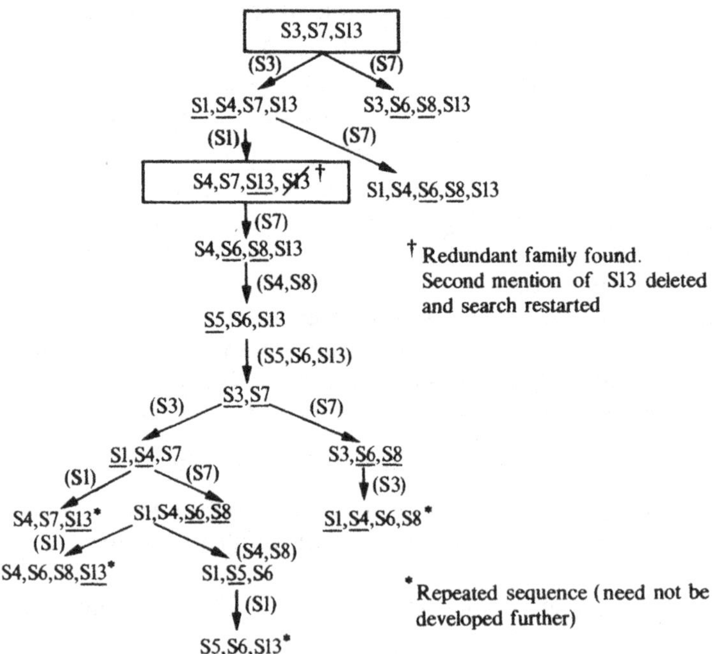

Figure 6.13. Applying the replacement rule of Upadhye and Grens to find the best tear set for group (CDEBA).

The next unit which would appear is unit A. We can define a stream string as a full string with the unit names deleted, and one can readily see that stream strings are one-to-one with full strings. An example derived from our first full string example is

$$(S1, S3, S5) \qquad \text{(stream string)}$$

They are more convenient to write down. A unit loop for a stream string (we are being quite careful here with nomenclature) is a stream string derived from a unit loop for a full string. Thus the following is an example:

$$(S1, S3, S13) \qquad \text{(unit loop for stream string)}$$

Note the unit loop is a *list of streams*, and it starts and terminates with the *same unit*.

We can also define stream loops for full or for stream strings. These loops start and terminate when the same stream is about to be repeated. An example is

$$(C, S5, D, S4, B, S3, C, S6, E, S7, C) \qquad \text{(stream loop for full string)}$$

where stream S5 is about to be repeated. Note this stream loop contains two unit loops.

The Upadyhe and Grens approach deals with unit loops for full (or stream) strings. We need first of all to list all unit loops for full strings for a flowsheet and such an algorithm is readily derived by extending the 'loopfinder' algorithm in Forder and Hutchison (1969). It consists of tracing all possible paths throughout a unit group. We illustrate with the group (ABCDE) in figure 6.11. We start with any unit and trace outputs, e.g. we start with unit A –

$$(A, S1, B, S3, (C, S5, D, S8, E, S7, 'C'))$$

We have found a unit loop: (C, S5, D, S8, E, S7). We look for an alternative output from the last unit, unit E (unit C does not count). There is none, so we look at the next to last unit, unit D:

$$(A, S1, B, S3, C, S5, D, \underline{S4, 'B'})$$

We have another unit loop: (B, S3, C, S5, D, S4). Again we look for another output from unit D. There is none so we back up to unit C, finding

$$(A, S1, B, S3, C, \underline{S6, E, S7, 'C'})$$

We have another unit loop: (C, S6, E, S7). No more outputs from E exist so we back up to unit C and find

$$(A, S1, B, S3, C, \underline{S13, 'A'})$$

which gives the loop (A, S1, B, S3, C, S13). C has no more outputs so we examine B (no more) and then A (no more). The loops found are indicated by the following loop/stream incidence matrix

	Stream	S1	S3	S4	S5	S6	S7	S8	S13
Loop									
1				1		1	1		
2			1	1	1				
3						1	1		
4		1	1						1

Motard and Westerberg noted that one can usually find a tear set for a flowsheet so each unit loop is torn exactly one time, and such a tear set belongs to the nonredundant family of Upadhye and Grens. Applying this idea, we proceed as follows.

Select a tear stream (stream S13 must be one, so select it): S13. Flag (and delete) all streams which are in any loops with S13. We flag streams S1 and S3 because they are in loop 4 with S13. We are left with

	Stream	S4	S5	S6	S7	S8
Loop						
1			1		1	1
2		1	1			
3				1	1	

Select another stream (it may as well be one that tears a maximum number of loops): S5. Again delete S5 and flag (and delete) streams S7, S8 (loop 1) and S4 (loop 2). This leaves us with

	Stream	S6
Loop		
3		1

S6 is our last tear stream and the tear set is (S13, S5, S6) which is indeed among the nonredundant family discovered earlier.

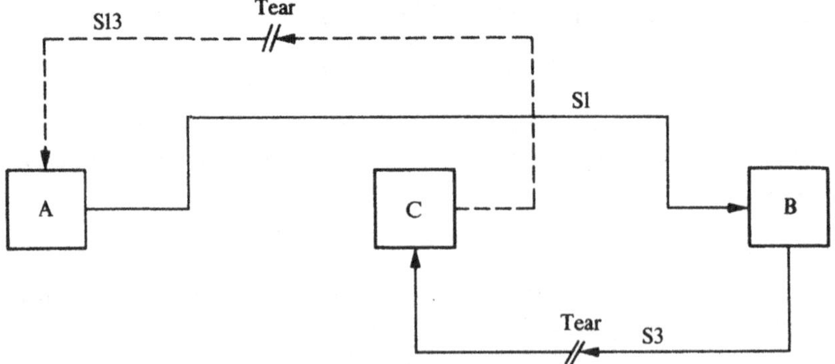

Figure 6.14. Loop 4 for example problem illustrating double tearing of loop.

Apparently then the nonredundancy goal of Upadhye and Grens is to avoid double tearing of unit loops for full (or stream) strings. Double tearing a unit loop puts a delay into it, as we can illustrate. Suppose the redundant tear set S3, S7 and S13 were used. We find loop 4 is double torn. The computational sequence for this tear set is (ACDEB) as illustrated in figure 6.12. Loop 4 is extracted and shown in figure 6.14.

We see the delay put into information flow quite clearly here. If the values for S13 are perturbed, we find the resulting perturbation does not make its way back to affect directly and immediately the calculated values for S13 via this loop. The perturbation makes it to unit B, but, since S3 is a feedback stream and is thus torn, the S13 perturbation does not pass from B to C via S3, even if successive substitution is used. It will take two passes through the computational sequence ACB, i.e.

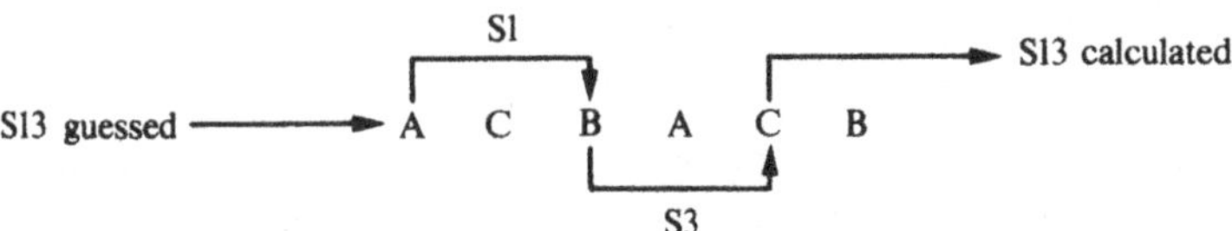

to get the effect of the perturbation to the calculated values for S13 via loop 4.

We can term a tear set an 'exclusive tear set' if it tears each unit loop exactly once for a group of units. Such a set does not always exist as the

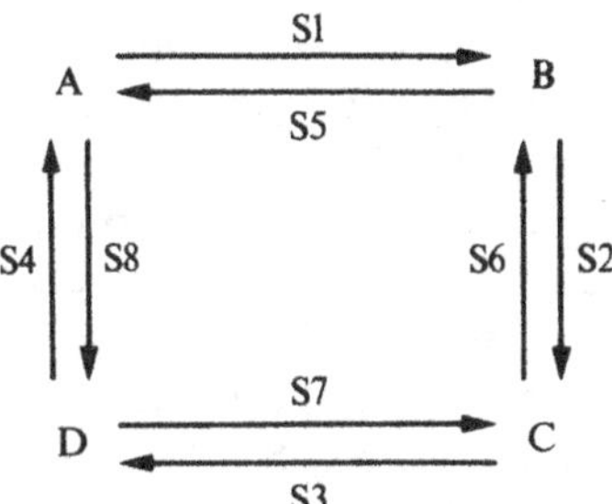

Figure 6.15. A group of units for which no exclusive tear set exists.

example in figure 6.15 illustrates. The loops for figure 6.15 are given by the following loop/stream incidence matrix:

Loops \ Stream	S1	S2	S3	S4	S5	S6	S7	S8
1	1				1			
2		1				1		
3			1				1	
4				1				1
5	1	1	1	1				
6						1	1	1

Selecting S1 as a tear stream flags and eliminates streams S2, S3, S4 (loop 5) and S5 (loop 1). We are left with the incidence matrix

$$
\begin{array}{c}
\text{Stream} \\
\text{Loops} \\
\begin{array}{c} 2 \\ 3 \\ 4 \\ 6 \end{array}
\end{array}
\begin{array}{ccc}
S6 & S7 & S8 \\
\left[\begin{array}{ccc} 1 & & \\ & 1 & \\ & & 1 \\ 1 & 1 & 1 \end{array}\right]
\end{array}
$$

Tearing stream S6 flags and eliminates the remaining streams, but loops 3 and 4 are never broken. A possible approach is then to recover loops 3 and 4 and all the streams involved.

$$
\begin{array}{c}
\text{Stream} \\
\text{Loops} \\
\begin{array}{c} 3 \\ 4 \end{array}
\end{array}
\begin{array}{cccc}
S3 & S4 & S7 & S8 \\
\left[\begin{array}{cccc} 1 & & 1 & \\ & 1 & & 1 \end{array}\right]
\end{array}
$$

We attempt to find an exclusive tear set for just this subproblem, knowing we must at least double tear other loops as we proceed. Selecting S3 tears loop 3 (and double tears loop 5). We can now select either S4 or S8 to tear loop 4, but S4 will triple tear loop 5 (loop 5 contains the tear streams S1, S3 already) whereas S8 will only double tear loop 6. We choose S8 to avoid the extra level of delay we would otherwise introduce into loop 5. Our tear set is (S1, S3, S6, S8).

Looking at figure 6.15, we see we have broken the loops (S1, S5), (S2, S6), (S3, S7) and (S4, S8) once each, the outer loop (S1, S2, S3, S4) twice and the inner loop (S8, S7, S6, S5) twice. This example contains, interestingly enough, three different nonredundant families of tear sets as would be discovered using the Upadhye and Grens definition and replacement rule. Each family is described as follows. The loops (S1, S5), (S2, S6), (S3, S7) and (S4, S8) are each broken once. Then family (1) breaks the outer loop (S1, S2, S3, S4) three times, the inner one once. Family (2) breaks the outer and inner loops twice each and family (3) breaks the inner loop three times, the outer once.

We can define the maximum of the number of times any loop is torn as the multiplicity of the tear set. Each tear set in families (1) and (3) has a multiplicity of 3 as for each (at least) one unit loop is torn three times. Each tear set in family (2) has a multiplicity of 2. Motard and Westerberg give a branch-and-bound algorithm for finding tear sets with the following attributes:

(*a*) Each stream can be given a user-defined weight. The higher the weight, the less desirable it is to tear the stream.

(*b*) No tear set of a lower multiplicity exists than the one found.

(*c*) Among the sets with this minimum multiplicity, the one with the lowest sum of weights is selected.

For the example in figure 6.15, a tear set from family (2) would be selected.

Extending the ideas we might wonder about the implication of using stream loops rather than unit loops. These loops might be more appropriate. Consider the cascade of units in figure 6.16. The unit loops are (S1, S6), (S2, S5), (S3, S4) whereas the stream loops (do not

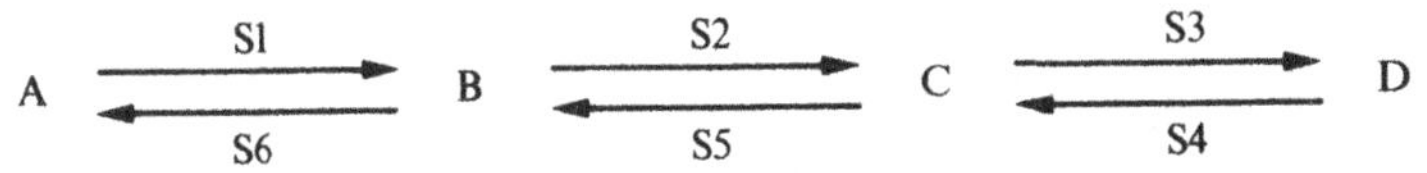

Figure 6.16. A cascade of units.

terminate a loop until a stream repeats) also include (S1, S2, S5, S6), (S2, S3, S4, S5) and (S1, S2, S3, S4, S5, S6). An exclusive tear set exists for the unit loops, e.g. S6, S5 and S4, with the computational order for the units being (ABCD). A perturbation in S6 in a single pass through the computation sequence can affect all the units by affecting S1, then S2 and then S3. However, this disturbance will not make its way via stream S4 back to unit C. Also the disturbance felt by unit C will itself not make its way back via S5 to unit B. We note the stream loops (S1, S2, S5, S6) and (S2, S3, S4, S5) are torn twice; the stream loop (S1, S2, S3, S4, S5, S6) is torn three times. Clearly then, we are forced to repeat the computational sequence (ABCD) three times to get the perturbation in S6 to return via the stream loop (S1, S2, S3, S4, S5, S6).

Observe the consequences of the computational sequence

$$(ABCDCB)$$

Note units C and B repeat but not in the same order. For this sequence all stream loops are broken exactly once only. No delay is experienced in getting a perturbation in S6 back via all stream loops. One might conjecture that this idea, to repeat the units if necessary in the computational sequence to have all stream loops torn but once, is valid and may lead to improved convergence behaviour. It has yet to be more fully investigated.

We might re-examine our earlier example in figure 6.12. Additional stream loops include

	Stream	S1	S3	S4	S5	S6	S7	S8	S13
Loop									
5			1	1	1	1	1		
6		1	1			1	1		1
7		1	1		1		1	1	1

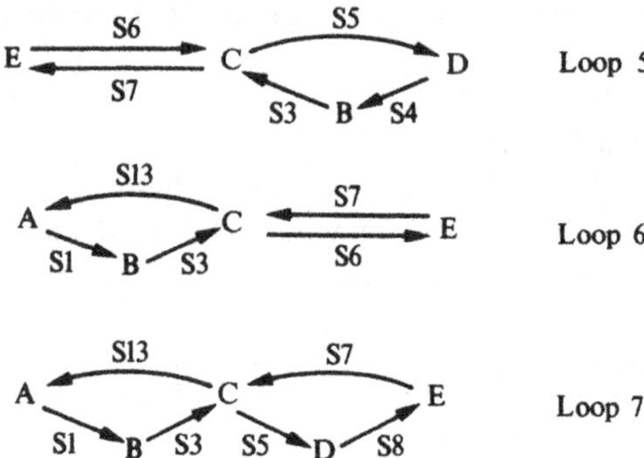

Figure 6.17. Three additional stream loops for unit group in figure 6.12.

They are each illustrated in figure 6.17.

Each has an obvious cascade appearance. The original computational sequence (ABCDE) would have to be repeated twice (each of these stream loops is doubly torn) to guarantee pertubations return via all stream loops. Alternatively one might use the sequence (ABCDECD). The original loops are torn once by the sequence (ABCDE). The new ones are torn once as follows:

Loop 5: B, S3, C, S7, E, S6, C, S5, D, S4   (BCECD)

Loop 6: A, S1, B, S3, C, S6, E, S7, C, S13   (ABCEC)

Loop 7: A, S1, B, S3, C, S5, D, S8, E, S7, C, S13   (ABCDEC)

Only units C and D need be calculated twice in the sequence. Alternative sequences include (ADBECDEC) for tear set (S5, S6, S13) and (EABCDEBC) for tear set (S4, S6, S8, S13). These may not be the best available but one can see alternatives exist; for example, one might prefer the last one over (ABCDECD) if units E, B and C are faster to re-evaluate than C and D.

7

Flowsheeting by equation-solving methods based on tearing

In this chapter we shall discuss the approaches possible to flowsheeting where one used tearing methods to deal directly with the underlying equation sets which define the units and the flowsheet. SPEED-UP (Leigh, Jackson and Sargent, 1974) represents a flowsheeting system of this type. The equations involved for a complex simulation are many, perhaps many times the estimate a novice is apt to make. The equations involved include

1. the user-supplied specifications,
2. the material balance equations,
3. the heat balance equations,
4. reaction equilibrium and/or stoichiometry equations and/or rate equations,
5. the equations connecting the units within the flowsheet,
6. correlations for heat, mass and momentum transfer,
7. and, perhaps in largest quantity, the physical property correlations.

This list is not intended to be all inclusive, so perhaps the reader can name even more equation types.

Such a flowsheeting system would require a library of unit models which would supply only the equations and *not* the method for their solution. The system would, from a description of the flowsheet, create the connection equations, and, if really general, the physical properties package could also supply only equations along with the appropriate parameters for the chemical species present.

After gathering all these equations together, the system would use algorithms such as those discussed in sections 3.4.1 and 3.4.2 to create a solution procedure for the equations. One could easily find this system is described by 10 000 or more equations, and if a large, multicomponent plate distillation column is present, the number could become 50 000 to 100 000 or more. Faced with such enormous numbers, compromises seem appropriate.

The first compromise is that it is likely that the physical properties package will not supply the equations only but will be in the form of a set of callable subroutines, just as for the conventional sequential modular approach. Using a very rough rule of thumb one might estimate that the equations involved in physical property evaluation could be four times all the other equations. This estimate results from some industrial experience which claims 80% of the time in a simulation is spent in the physical properties package (P. T. Shannon, personal communication). Counting equations for a flash calculation using Chao–Seader correlations indicates this ratio is not far wrong.

A second compromise may be to use this approach together with simple models only. Thus it would be used early on in the design process only. Its enormous advantage of allowing very arbitrary user specifications to be handled in a direct manner is needed in these early calculations.

Another strategy is also valuable, that of approaching the problem at more than one level of analysis. For example, one can first analyse the information flow for a problem at the flowsheet level to ascertain the potential for having a significant impact on computation time by carefully designing the information flow at that level. One may discover, for example, that the apparent number of recycle calculations at the flowsheet level can be reduced or even eliminated.

In this chapter we shall explore the use of analysing the information flow for a flowsheet at two levels. Although the natural result might then appear to yield two levels of iteration for solving the equations, we shall also discuss an apparently better strategy: derive the solution procedure at more than one level, but converge the resulting iterations from all levels only at the outer level.

7.1 A simple example

The information flow for a unit can be indicated by directed arrows. Figure 7.1 shows several different flows for information for a reactor unit where $c = 5$ and $r = 2$ (c = number of components and r = number of reactions). In chapter 5 we established that a stream with $c = 5$ components has $c + 2 = 7$ degrees of freedom (i.e. seven independent variables) associated with it. Also for our simple reactor model with two independent reactions occurring, we had that $r + 2 = 4$ equipment parameters existed and that the unit represented $c + 2 = 7$ equations among the input and output stream variables and equipment parameters. We did not establish just *which* variables the unit equations would calculate given values for the others. An arrow into the block will be used to indicate we know as input a value for the number of variables indicated, and we shall use this block to calculate the values for the variables represented by the arrows pointed away from the block. The total number of the calculated variables (i.e. those associated with outward pointing arrows) must *equal* the number of equations in the block.

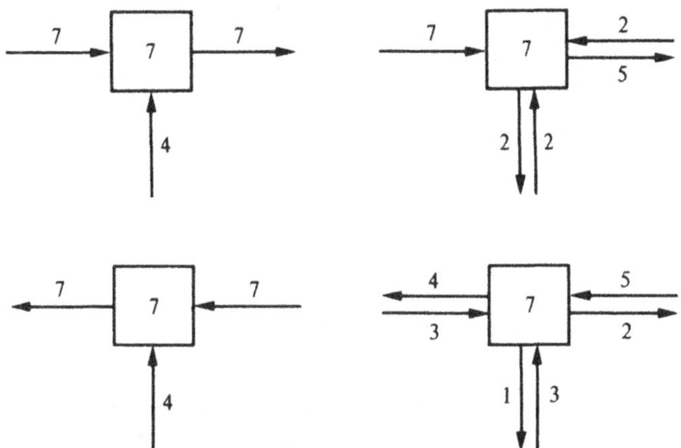

Figure 7.1. Various possible flows of information in a reactor unit with $c = 5$ components and $r = 2$ independent reactions.

In figure 7.1 the first diagram indicates that we shall solve for the output stream values given the input stream and equipment parameter values. The top right diagram indicates that we intend to calculate only five of the output stream values plus two of the equipment parameters. We shall be given the input stream parameters, two equipment parameters, and two output stream variable values. The remaining two diagrams illustrate other variations for information flows.

If the unit calculations prove to be very time consuming, we would like to choose which variables should be the decision variables to our best advantage. It would be useful, for example, if we could eliminate all recycles in a flowsheet. Consider the very simple block diagram in figure 7.2(a). The conventional sequential modular procedure would result in the computations implied by figure 7.2(b) where the seven variables in the recycle stream are guessed and iterated. If in fact we have no *a priori* preference for which variables we choose as decisions, the arrangement in figure 7.2(c) has eliminated all recycle variables at the flowsheet level of computation by a different choice as to which variables are decision variables. We use the 'recycle' variables as decisions here and calculate the flash, reactor and mixer units backwards, in each instance, to get their inputs.

It is questionable whether one would have a problem where all the feed stream values are unspecified, but the example proves that the potential is there for reducing the number of recycle variables significantly.

It should be noted again that we are working on an hierarchical view of the problem. In figure 7.2, the equations within each unit are treated as though they are to be solved *together*. The exact solution procedure

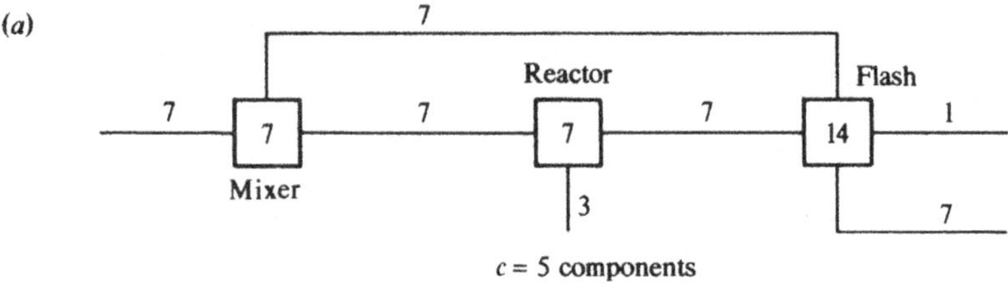

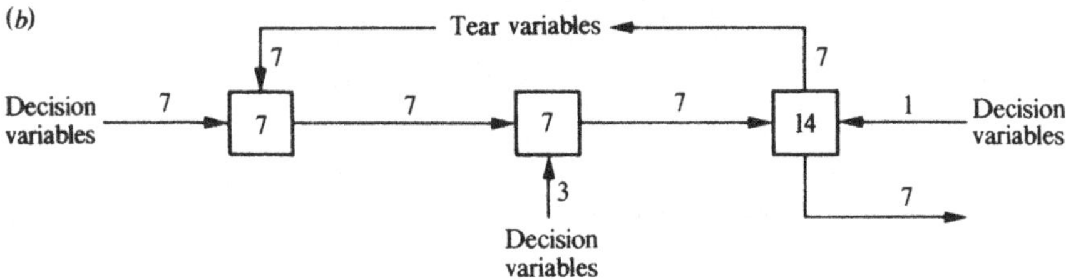

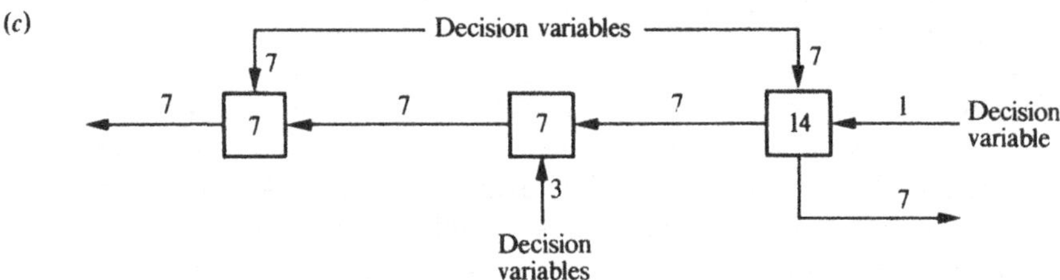

Figure 7.2. A simple example illustrating the potential for choosing the decision variables in a flowsheet to reduce computational effort.

for each unit will depend on the information flow at the flowsheet level. We may, by eliminating recycles for the flowsheet, force more recycle calculations within the unit computations. One tends to prefer to have several small iterations to one large one, so this approach should be appealing. In figure 7.2(c) we know we have to solve each unit exactly once and we are done, even if internally we need some extra iterations for each unit. If the potential for recycles within the units is troublesome, we can, in principle, gather all the equations together for some, or all, of the units as one set, and develop a combined solution procedure for them.

The hierarchical view has another advantage. We can develop a single procedure for the flowsheet level of the problem; then the unit models can be changed as more exact modelling is required. This advantage may be a marginal one.

Lastly, if one is solving a large flowsheet, one may discover that the feed preparation units can be analysed separately from the reaction section, and these both can be analysed separately from the product purification section by cleverly selecting the decision variable. This result means three parallel design efforts can take place, and here the view is extremely valuable. Figure 7.3 illustrates this use. In this

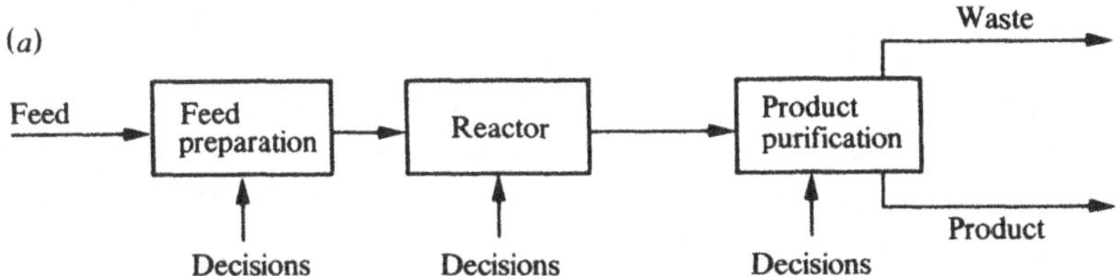

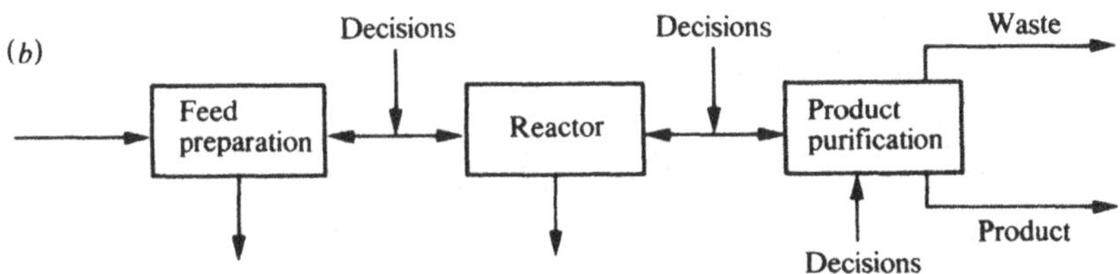

Figure 7.3. Choosing decisions to partition a flowsheet. In (*a*) the sections must be calculated in sequence; in (*b*) the three can be calculated in parallel.

example, one decides the characteristics for the stream which flows from the feed preparation system and into the reactor system and also for the reactor output stream. This latter stream feeds the purification section. Having selected these stream specifications, one can assign one design team to design a feed preparation system capable of converting the given plant feed into the specified reactor feed. In parallel, since the reactor feed is known, a second design team can design a reactor system to convert the reactor feed to the specified product purification feed; this last system can be designed in a third parallel effort. In fact this type of partitioning is routinely done.

7.2 An example system based on equation solving

Westerberg and Shah (1978) describe a programming system developed to analyse and optimize simple heat exchanger networks, adjusting exchanger areas, temperature levels, and/or flows to give a network with the best annual cost.

The general approach taken is to model each unit in such a network functionally, by writing overall material and energy balances. One creates few equations per unit in this manner. Using the functional equations, the modelling of such a system is essentially at the outer (i.e. flowsheet) level, and a solution procedure is always sought to eliminate, if possible, computational recycles at this level. The unit models themselves can, in fact, be very complex depending on the detail sought. They have to be written so any allowed information flow required at the outer level can be satisfied.

The above approach has been taken because the modelling equations are being used when optimizing the given network. Thus the equations are solved repeatedly as an inner loop to an optimization program. Figure 7.4 illustrates the structure. The optimizer block directs all the

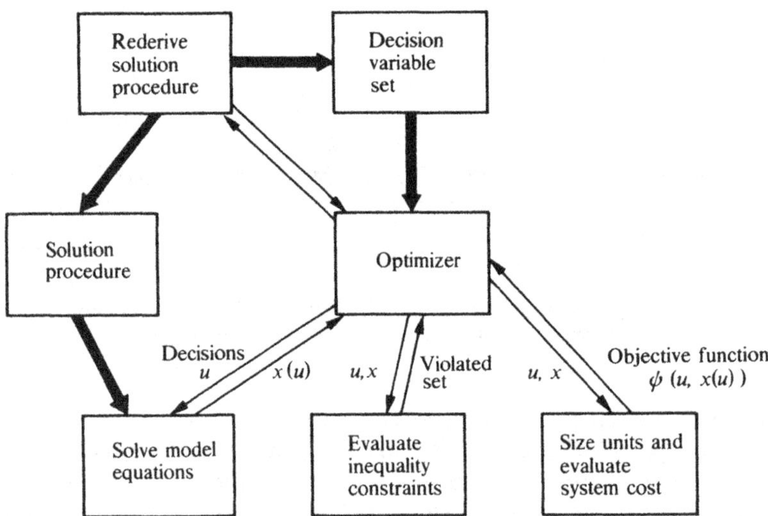

Figure 7.4. Structure of an optimization system. Bold arrows indicate transfer of information; normal arrows indicate transfer of control.

activity here. It is adjusting the decision variable values to improve ψ, the objective. For this system ψ is the 'annualized' cost of the equipment plus the annual cost of buying the utilities needed, such as steam for heating, and water for cooling.

To evaluate ψ, the optimizer supplies the block labelled 'solve model equations' with the values it wishes to try for the decision variables, u. Solving the model equations gives values for the remaining problem variables (e.g. flows, temperatures) which we shall call $x(u)$. With u and $x(u)$ values available, constraint violations are checked, and, if some are violated, they are identified to the optimizer. Assuming none are violated, the units in the system are sized and a cost is evaluated. The optimizer notes this cost and changes the decision variable values, in the hope of reducing the resulting objective function. This calculation sequence is repeated many times during a typical optimization. If constraints are violated, special action is taken, which for this system will result in a modified set of model equations and the need to rederive a solution procedure for them (this approach is based on the optimization algorithms in deBrosse and Westerberg (1973) and Westerberg and deBrosse (1973)).

The need for efficient model calculations is evident. One could be required to solve the model equations several hundred times, and an unnecessary computational loop is no longer a tolerable burden. The inequality constraints are large in number, and the system, in practice, will need to modify the equation set about as many times as there are inequality constraints present in the search for the right set to be held at the optimum.

In this chapter we shall describe a simplified version of the above system, starting first with the modelling considerations. Then we shall describe the needed data input for a small sample problem. Lastly, we shall describe one pass through preparing the problem for solution.

7.2.1 Modelling considerations

The model for a network of heat exchangers is discussed here. We shall consider four aspects of the model: the flowsheet, the streams and stream segments and lastly the unit types. Figure 7.5 is an example network.

The network comprises a single hot process stream H1 which is split and used to heat two cold process streams, C1 and C2. It then merges and is cooled to its exit conditions by a cooling water stream, W. Streams C1 and C2 are preheated by steam utility streams S1 and S2. We note the network has five heat exchanger units, A, B, E, F and G, one stream splitter unit, C, and one mixing unit, D. The streams have been broken up into segments, of which we have 19 overall. For example, stream C1 enters unit E as segment 8. It exits and proceeds to unit A as segment 9, and finally leaves the system as segment 10. Hot stream H1 enters as segment 1. It is split into two segments, 2 and 3, by the splitter unit C, and so forth. The naming scheme should now be evident.

The streams are C1, C2, H1, S1, S2 and W for our problem. We need to specify, or have calculated, for a stream the following items:

1. Flowrate F_k, $k = $ C1, C2, H1, S2, S2, or W
2. Stream inlet temperature $T_{k,\,\text{in}}$ and vapour fraction $\varphi_{k,\,\text{in}}$

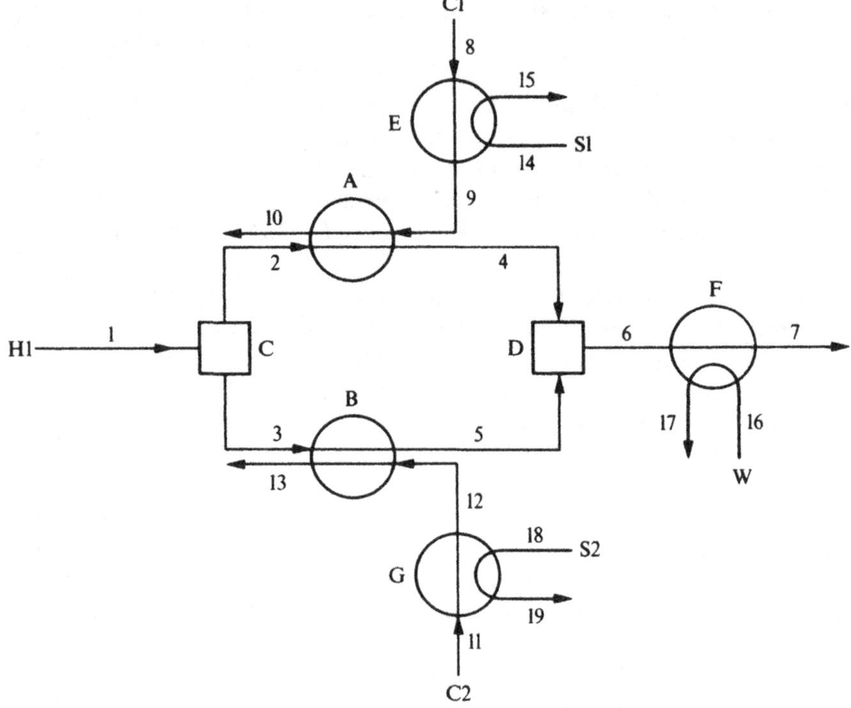

Figure 7.5. An example heat exchanger network.

3. Stream outlet temperature $T_{k,\,out}$ and vapour fraction $\varphi_{k,\,out}$
4. Stream cost data (process streams are usually considered as free sources of, or sinks for, energy; utility streams, on the other hand, have a cost associated with them)
5. Stream physical property data
6. Stream materials-of-construction data

Each of the items 1 to 3 may be specified by supplying a given value or only a set of bounds. For example, a process stream may be specified by stating

1. $F = 10$ kg/s
2. $T_{in} = 350$ K, $\varphi_{in} = 0$ (liquid)
3. 460 K $\leqslant T_{out} \leqslant 475$ K, $\varphi_{out} = 1$ (vapour)

and a cooling water utility stream by stating

1. $0 \leqslant F \leqslant 100$ kg/s
2. $T_{in} = 300$ K ($80\,°F$), $\varphi_{in} = 0$
3. $T_{out} \leqslant 325$ K ($125\,°F$), $\varphi_{out} = 0$

The stream segments can be characterized by just four parameters if we assume all segments for a stream are of the same composition and generally the same pressure. The parameters are

1. k stream identifier (the segment is part of stream k)
2. F_i flowrate
3. h_i enthalpy/unit flow
4.   vapour fraction of stream

We could add a pressure parameter in a more accurate simulation system, but we wish to keep things moderately simple here.

The unit models are written functionally by writing overall material and energy balances. Also, we can write associated inequality constraints, equipment sizing equations, and finally how to evaluate the cost for a piece of equipment. We shall treat each unit type in turn.

Heat exchanger

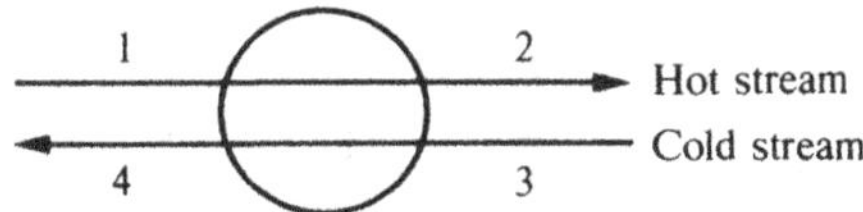

Four stream segments connect to a heat exchanger. We can write two material balances and one overall energy balance for this unit.

Material balance:

$$F_2 = F_1$$
$$F_4 = F_3$$

Energy balance:

$$h_1F_1 + h_3F_3 = h_2F_2 + h_4F_4$$

The basic inequality constraint is that at no point in the exchanger should the hot stream temperature be equal to, or fall below, that of the cold stream. Usually this desire is written as

$$T_1 \geqslant T_4 + \varepsilon$$

$$T_2 \geqslant T_3 + \varepsilon$$

where ε has a small positive value, say 2K.

However the temperature could cross over internally and these constraints could easily fail to detect it, particularly when a stream undergoes a phase change. One needs to check, therefore, at several points along the exchanger to prevent 'crossover' temperatures. Also we would like the hot stream to be cooled (and thus the cold stream to be heated) when passing through the exchanger. We include, therefore, the constraint.

$$T_2 \leqslant T_1$$

The sizing calculation for an exchanger is to evaluate its area. This calculation can be very involved, but for design purposes may, in fact, be simplified by assuming film coefficients based on the fluid, whether it is heating or cooling, and whether it is boiling or condensing (see Perry and Chilton, 1973). The exchanger may, again for simplified design purposes, be considered to operate in zones, as indicated in figure 7.6. Each zone is then sized using the appropriate film coefficients and temperature driving forces. T_D and T_B indicate dew and bubble points for the streams.

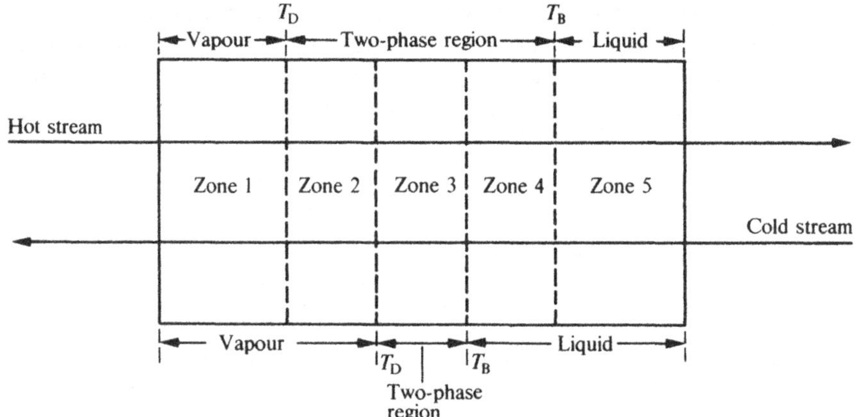

Figure 7.6. Partitioning an exchanger into zones when phase changes occur.

The checking for crossover temperatures is significantly improved if it is done at the boundaries of the zones as well as at the exchanger endpoints.

To place a crude cost on the exchanger, one might use an equation of the form (see Guthrie, 1969)

$$\text{Cost} = f_m f_P (aA^m)$$

where f_m is a materials factor, f_P a pressure factor and the term aA^m is the cost of a carbon-steel exchanger operating at one atmosphere. The terms a and m are constants, with m being about 0.6 to 0.8.

Stream splitter unit

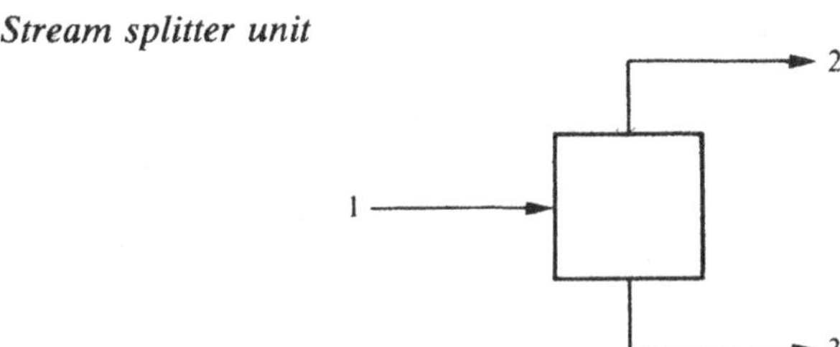

A stream splitter unit simply divides a stream into two parts, leaving all characteristics except the flowrate unaltered. The material and energy balances are as follows.

Material balance:

$$F_2 = aF_1$$
$$F_3 = (1-a)F_1$$

Energy balance:

$$h_2 = h_1$$
$$h_3 = h_1$$

The only inequality constraints needed are on a, the split factor,

$$0 \leqslant a \leqslant 1$$

We shall assume a splitter unit will cost us a constant amount of money – that necessary to buy and maintain the flow equipment needed to control the flow splitting.

Mixer unit

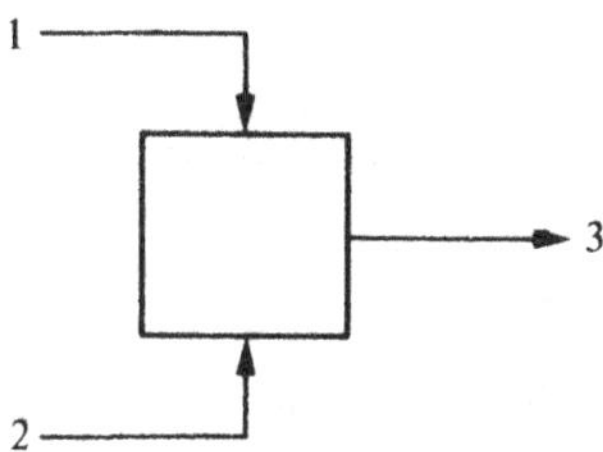

The mixer unit merges two stream segments into one and has the following balances.

Material balance:

$$F_3 = F_1 + F_2$$
$$h_3 F_3 = h_1 F_1 + h_2 F_2$$

No constraints need be written and the unit can be assumed to have no associated cost.

The last source of equations is the evaluation of physical properties. The system must be able to convert from stream temperature (and vapour fraction, if a pure component is in the two-phase region) to enthalpy and back again. One could enter in some form the cooling curve for the stream if one assumes the stream is at essentially a constant pressure. Figure 7.7 illustrates a cooling curve, where T_D and T_B are dew and bubble point temperature respectively.

Also one will need thermal conductivities, densities and so forth if correlations are to be used when evaluating film coefficients. If for design purposes typical values for film coefficients are used, these properties will not be needed.

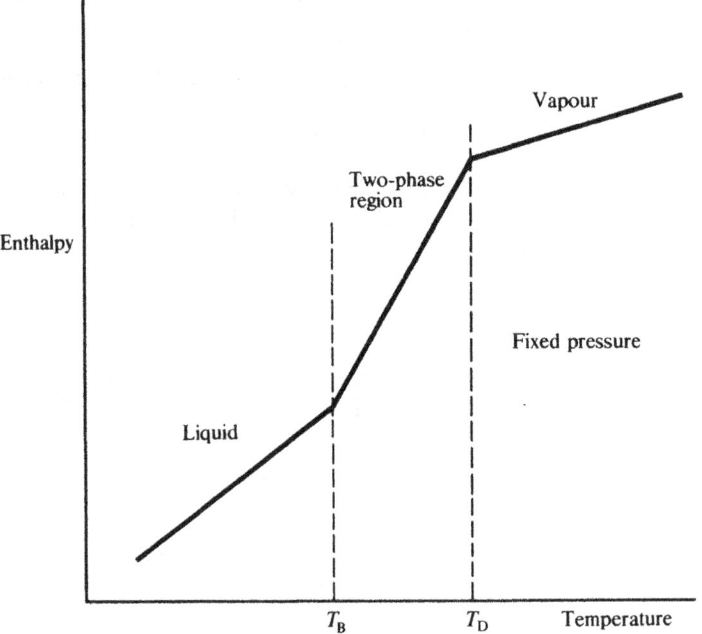

Figure 7.7. A typical cooling curve.

7.2.2 Data specification for example problem

We need to supply enough data into the computer to define our problem. A problem definition will require that we input

1. the flowsheet structure: connectivity of segments and units
2. the unit data
    (*a*) unit type
    (*b*) any equipment parameter specifications

3. the stream data
    (*a*) flow and temperature specifications
    (*b*) physical property data
    (*c*) film heat transfer coefficients
    (*d*) materials specifications and other cost parameters
4. the segment data
    (*a*) associated stream identifier
    (*b*) any specifications imposed on flow and temperature
5. general user specifications
6. guessed set of inequality constraints to be held

The following is typical of the type of data input we might use.

(1)   *Flowsheet (see figure 7.5)*

Segment no.	Stream identifier	Source unit	Destination unit
1	H1	Outside	C
2	H1	C	A
3	H1	C	B
4	H1	A	D
⋮			
8	C1	Outside	E
9	C1	E	A
⋮			
19	S2	G	Outside

(2) *Unit data*

Unit name	Unit type	Parameter	Specification LB	UB	IV
A	Heat exchanger				
B	Heat exchanger				
C	Splitter	*a*	0	1	0.5
D	Mixer				
E	Heat exchanger				
F	Heat exchanger				
G	Heat exchanger				

LB is lower bound, UB is upper bound and IV is initial value. If LB and
UB are both not specified, IV becomes the fixed value for the variable.
We shall use an asterisk to indicate that the value is not specified.

(3) *Stream data*

Stream		H1	W	etc.
Flowrate	LB	*	0	
(kg/s)	UB	*	100	
	IV	10	5	
T_{in}	LB	*	*	
(K)	UB	*	*	
	IV	450	300	
φ_{in}	LB	*	*	
	UB	*	*	
	IV	1	0	
T_{out}	LB	250	*	
(K)	UB	275	325	
	IV	275	325	
φ_{out}	LB	*	*	
	UB	*	*	
	IV	0	0	
Cooling curve data h_L h_{TP} h_V		⋮	⋮	
Materials data		⋮	⋮	
Cost	($/kg)	0	7×10^{-5}	

(4) *Segment data*

The associated stream is already identified for each segment in the flowsheet data section. Thus only those segments with specifications on them need be listed here.

Segment Number	F			T		
	LB	UB	IV	LB	UB	IV
4	–	–	–	250	*	300
5	–	–	–	225	*	300
6	–	–	–	250	*	300
9	–	–	–	*	300	225
12	–	–	–	*	325	225

We are using this data entry to establish first guesses for the intermediate temperatures to aid (we hope) in starting the problem.

(5) *General user specifications*

We may wish that a ratio be maintained between two flows; such a specification could be entered here. The types of specifications allowed will obviously depend on the design of the input language for the system.

(6) *Guessed set of inequality constraints to be held*

The user may have an idea of which constraints, if held, would yield the optimal system operation. If so he could enter their identifiers here. For example, he may believe that, at the optimal solution, the exit temperature of H1 in exchanger A will be at the allowed 2 K approach temperature above the inlet cold temperature for C1. Also he may believe that the steam utility stream S2 should not be flowing, i.e. exchanger G is not really needed. He could specify that these two inequality constraints be included in the original solution procedure derived. Again the mechanism used to identify and specify these constraints is a function of the design of the input language for the system.

7.2.3 Deriving a solution procedure and solving

We shall now consider the steps needed to develop a solution procedure and then to solve it for our example problem. First the system must gather together the necessary equations, or at least establish their structure, so a solution procedure may be prepared. The desired solution procedure should eliminate all recycle loops in the computations, if possible, or minimize their number if not possible.

The initial solution procedure will ignore all but the inequality constraints specified in part (6) of the last section. Thus, we will need to gather together the heat and material balance equations for each unit in the process and the fixed variable specifications. These equations include the following:

Unit

A
$$F_4 = F_2 \tag{7.1}$$
$$F_{10} = F_9 \tag{7.2}$$
$$h_2F_2 + h_9F_9 = h_4F_4 + h_{10}F_{10} \tag{7.3}$$

B
$$F_5 = F_3 \tag{7.4}$$
$$F_{13} = F_{12} \tag{7.5}$$
$$h_3F_3 + h_{12}F_{12} = h_5F_5 + h_{13}F_{13} \tag{7.6}$$

C
$$F_2 = a_C F_1 \tag{7.7}$$
$$F_3 = (1 - a_C)F_1 \tag{7.8}$$

$$h_2 = h_1 \tag{7.9}$$

$$h_3 = h_1 \tag{7.10}$$

D
$$F_6 = F_4 + F_5 \tag{7.11}$$

$$h_6 F_6 = h_4 F_4 + h_5 F_5 \tag{7.12}$$

E
$$F_{15} = F_{14} \tag{7.13}$$

$$F_9 = F_8 \tag{7.14}$$

$$h_{14} F_{14} + h_8 F_8 = h_{15} F_{15} + h_9 F_9 \tag{7.15}$$

F
$$F_7 = F_6 \tag{7.16}$$

$$F_{17} = F_{16} \tag{7.17}$$

$$H_6 F_6 + h_{16} F_{16} = h_7 F_7 + h_{17} F_{17} \tag{7.18}$$

G
$$F_{19} = F_{18} \tag{7.19}$$

$$F_{12} = F_{11} \tag{7.20}$$

$$h_{18} F_{18} + h_{11} F_{11} = h_{19} F_{19} + h_{12} F_{12} \tag{7.21}$$

Let us assume that the user has requested that we include the constraint stating the exit temperature for stream H1 is at the approach temperature allowed. We do this as follows.

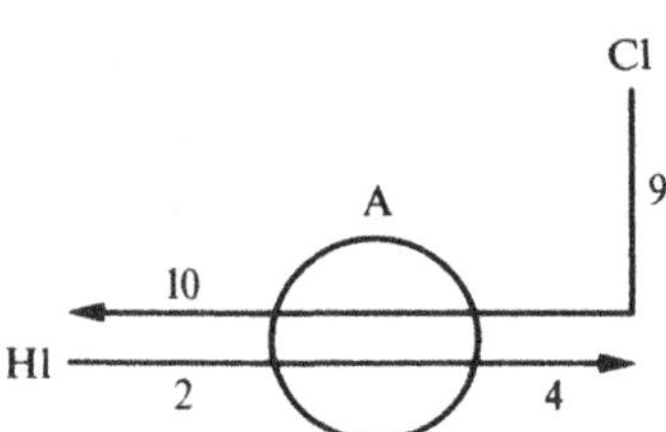

The inequality constraint is converted to an equality constraint by introducing a slack variable σ.

$$T_4 = T_9 + \varepsilon + \sigma_{49} \qquad \sigma_{49} \geqslant 0 \tag{7.22}$$

where ε is the allowed approach temperature, say 2 K. The slack variable is required to be nonnegative. When we derive the solution procedure, we shall require σ_{49} be a decision variable with an initial value of zero for the problem. In this way, we will force $T_4 = T_9 + \varepsilon$ initially.

We note that this constraint is in terms of temperature rather than enthalpies. We need to add, therefore, the relationships

$$T_4 = f(h_4) \tag{7.23}$$

$$T_9 = f(h_9) \tag{7.24}$$

to our equation set.

Where are our connection equations? The system has implicitly accounted for them by using the appropriate segment variables when constructing the model equations. For example, F_{12} appears in the equations for both units B and G; segment 12 connects these two units.

We are now in a position to set up an incidence matrix. We can significantly reduce its size by a bit of cleverness on our part. We note that a large percentage of the equations written are simply equating one variable to another. Equations (7.1), (7.2), (7.4), (7.5), (7.9), (7.10), (7.13), (7.14), (7.16), (7.17), (7.19) and (7.20) are precisely of this form. These equations will automatically be satisfied if we store the value of any variables so equated in a common storage location. We can therefore delete these equations and merge the variables, giving us the incidence matrix in figure 7.8.

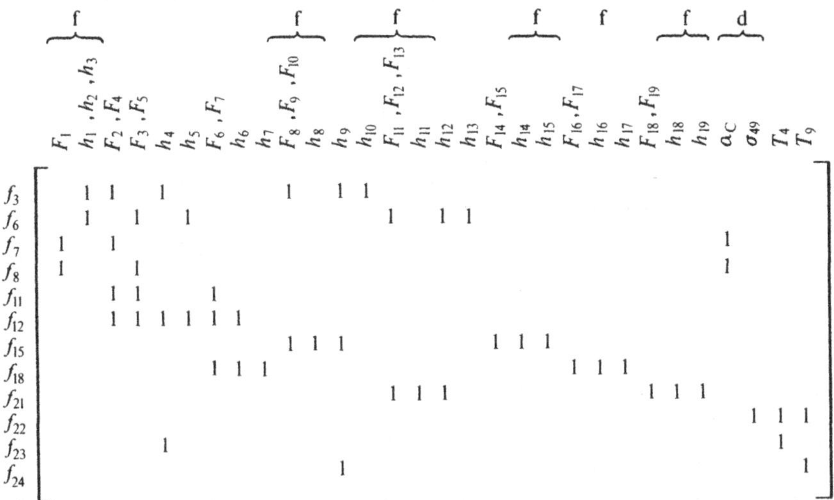

Figure 7.8. Incidence matrix for example problem. d, decision variables; f, variables with fixed input values.

Many of the variables in the incidence matrix are, in fact, specified. For example, in our data input in the previous section, we directly, or indirectly, specified the following:

F_1 Flowrate for H1
h_1 T_{in} and φ_{in} for H1
h_{16} Cooling water T_{in} and φ_{in}

We might assume, if we had completed the stream data table, the following additional specifications would be there too.

$h_{14}, h_{15}, h_{18}, h_{19}$ The entry and exit stream thermal conditions specified for S1 and S2

F_8, h_8, h_{10} — Entry and exit conditions specified for C1

F_{11}, h_{11} — Entry conditions for C2 specified (the exit temperature is bounded)

These columns are marked with an 'f' indicating that the variables they represent have fixed input values.

We also stated earlier that σ_{49} was a required decision variable with a prescribed initial value of zero. We shall, since a_C is bounded between zero and one, also require a_C to be a decision variable. These two columns are labelled 'd'. We delete all columns labelled 'f' and 'd' getting figure 7.9.

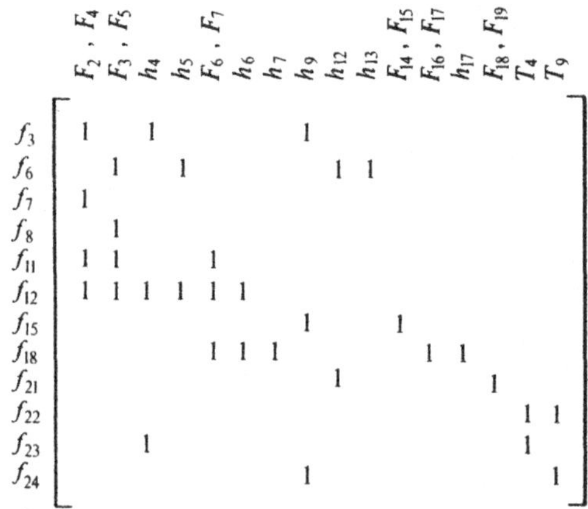

Figure 7.9. Reduced incidence matrix after all prespecified and design variable columns have been deleted.

We can now apply the algorithm described in section 3.4.2 to find the following solution procedure.

1. Decisions $h_5, h_7, h_{17}, F_{18} \ (= F_{19})$
2. f_7 calculates $F_2 \ (= F_4)$
3. f_8 calculates $F_3 \ (= F_5)$
4. f_{11} calculates $F_6 \ (= F_7)$
5. Guess h_4
6. f_{23} calculates T_4
7. f_{22} calculates T_9
8. f_{24} calculates h_9
9. f_3 calculates new h_4
10. If h_4 (guessed) essentially equal to h_4 (calculated), continue with step 11. Otherwise reguess h_4 and iterate from step 6
11. f_{12} calculates h_6
12. f_{12} calculates h_{12}

13. f_6 calculates h_{13}
14. f_{18} calculates F_{16} $(= F_{17})$
15. f_{15} calculates F_{14} $(= F_{15})$

We see we have an iteration loop involving the single tear variable h_4 in the middle of this calculation. The loop is explicit in h_4, in that step 9 above calculates a value for h_4 rather than simply evaluates an error function whose value is only implicit in h_4.

The above solution procedure now fits into the block labelled 'solve model equations' in figure 7.4, the figure showing the general structure of the optimization system. The optimizer has a selected set of six decision variables to optimize: $a_C, \sigma_{49}, h_5, h_7, h_{17}$ and F_{18} $(= F_{19})$. There is a very high probability that the optimizer will manipulate the decision variable values, causing a number of constraints to be violated. Each in turn can be introduced with a slack variable; the slack will be pre-selected to be a decision variable in the derived solution procedure. In this way the problem will in all likelihood reduce to one with all but one or two (or zero) of the variables being slack variables. The slack variables will, if the constraint should be held, be set to zero and will not be manipulated. Thus the problem will turn into an optimization problem involving only one or two actual degrees of freedom, and perhaps none.

We have said nothing about starting the problem where our starting point is very likely infeasible. An algorithm is available (deBrosse and Westerberg, 1973) which selectively introduces and deletes constraints until a set leads to a feasible point or can be used to show no feasible point exists. Introducing and deleting constraints again requires one to rederive the solution procedure with each change.

The final and important point remains on just how the model equations have to be written. We must write for a heat exchanger unit a routine which can solve the heat balance equation in an arbitrary fashion.

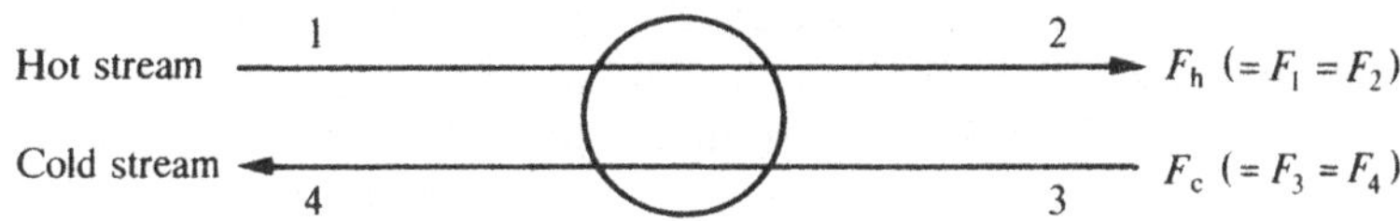

$$h_1 F_h + h_3 F_c = h_2 F_h + h_4 F_c$$

The routine must be written to permit any one variable to be calculated in terms of the remaining ones. This requirement is straightforward, for example

$$F_h = \frac{F_c(h_4 - h_3)}{h_1 - h_2}$$

unless the denominator $h_1 - h_2$ becomes zero. To preclude such problems, one can always force all flows to exist; i.e. all flows must satisfy

$$F \geqslant \delta > 0$$

where δ is a small number. If no flows can be zero, we avoid enthalpy changes which are zero, and so we effectively avoid the dividing-by-zero problems otherwise possible.

For a splitter unit, we can stop a zero flow from leaving by requiring

$$0 < \delta \leqslant a \leqslant 1 - \delta < 1$$

These precautions will make it easy to write the needed routines, which incidentally can range from the very simple to very complex depending on the modelling sophistication used. If very complex, then our strategy to minimize recycle computation loops at the flowsheet level pays off handsomely.

7.3 A complex example of selecting decision and tear variables for a flowsheet

In chapter 5, we developed an example flowsheet comprising a feed stream mixing with a recycle stream and becoming a feed to a reactor. The reactor effluent was flashed to separate off a liquid product stream and to recycle unreacted feed back as the vapour stream. A bleed was included in the vapour recycle to reduce build up of an inert contaminant entering in the feed gas. Figure 5.12 illustrates this example. Figure 5.13 then gives its basic block diagram, which we used to determine the degrees of freedom for this example. In chapter 6 we solved a particular case of this problem using the sequential modular approach.

We would now like to examine, at the flowsheet level, the potential available to reduce the number of required recycle computations. The ideas included here will, in fact, extend those given in section 3.4.2 on finding a good solution procedure when faced with a problem described by more variables than .equations. The new twist will be our ability to handle logically a function which will calculate the values for several variables simultaneously; a computer subroutine corresponds to this type of function. Until now, we have dealt with functions that can calculate exactly one output variable; even the heat exchanger system in the previous section had this property.

The approach envisaged here is the hierarchical one mentioned in the introduction to chapter 7. First we look at the flowsheet level to reduce iterations. The result will be an information flow at this level which will force us to look then at the unit equation sets to see if they can be solved in the manner which is required; that is, can they be written to receive, say, five or seven of the variables in the top stream as input and calculate all of the feed stream variables and all but three of the remaining output stream and equipment parameter variables? The three not calculated will be, again, input to the calculation.

If we cannot satisfy the required information flow, an iteration results where we must revise our information flow at the flowsheet level. We will not, with this example, run into this iteration. We shall try to motivate the ideas as we develop and use them.

Consider again the system illustrated by figure 5.12. Suppose we were going to solve it by hand; how should we do it with somewhere near the least effort possible? Our goal is to show how the step of deriving a solution procedure is amenable to being automated. Again it will be very similar to the ideas developed earlier for solving nonlinear equations using tearing methods. For a small flowsheet the basic block diagram introduced in chapter 5 can be used, but, for a larger one, a method using an incidence matrix representation would prove to be less tedious. We shall examine the incidence matrix formulation later.

For this flowsheet, we shall assume we already have in hand, or require, the following specifications:

		No. of specifications
1.	The feed composition and pressure	2
2.	Inlet T to reactor	1
3.	The reactor is adiabatic ($Q = 0$)	1
4.	Cooling water T_{in} and P_{in}, and T_{out}	3
5.	Product purity (mole fraction of component C) and total flowrate out of flash unit	2
6.	Fractional recovery of C in flash (ratio of C leaving product to that in feed)	1
7.	The flash is adiabatic	1
8.	Efficiency of compressor and outlet pressure	2
	Total no. of specifications	13

Figure 7.10 contains the resulting changes and information flows immediately evident from these specifications. Item (6) above has led to some added blocks (R1, R2 and R3) since it is an equation involving the ratio of the amount of component C in the product to that in the feed for the flash unit, and is of the form

$$\text{Specified ratio} = (x_C F)_{\text{product}}/(x_C F)_{\text{feed}}$$

See the basic block diagram for a user specification given in section 5.3. Block R1 provides seven equations: five transfer the five input variables to the five output variables, and two transmit the values of x_c and F to R2. Similarly R3 is equivalent to seven equations, five of which state that the five variables on the topside and bottomside are equal, and two

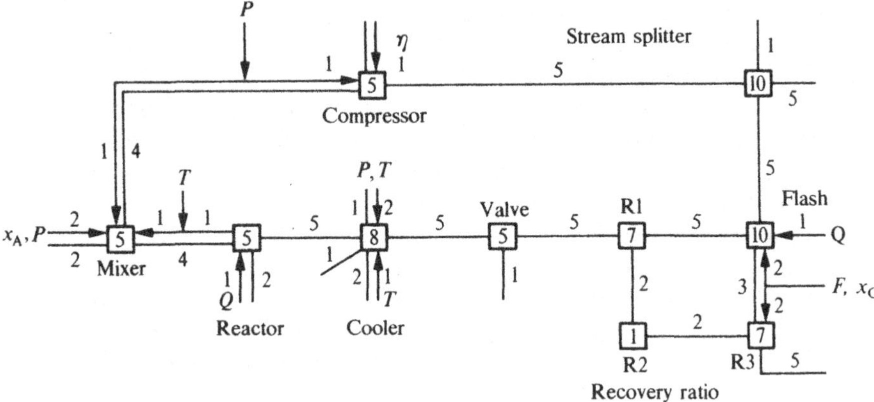

Figure 7.10. Basic block diagram with design specifications added.

of which are needed to transfer the values of x_c and F in the product to R2. R2 receives four variables, x_c and F in the flash feed and x_c and F in the flash product; it represents one equation in these four variables. We note that three degrees of freedom still remain at our disposal.

To derive a solution procedure we proceed as follows.

A. Find any units with an equal number of undirected variables as equations. Clearly, these equations can be used immediately to calculate those variables. Delete the unit and indicate that its equations should be used next. Repeat for any other units found.

   The mixer has five equations and 10 unassigned variables, the reactor five and 11, the exchanger eight and 14, the valve five and 11, and so forth. None satisfy this rule. Go on to rule B.

(This step corresponds to finding a single equation which involves a single remaining unknown variable. It is clear that we can usually solve such an equation immediately for the value of the remaining variable.)

B. Do any unassigned variables appear which are connected to only one unit? If so they must be calculated by the equations for that unit or they must be made decision variables. Because they do not connect to any other equations, they cannot be inside a recycle loop of calculations if we require them to be calculated. This should be evident since, after being calculated, they are not used anywhere else in another calculation.

(Again this step corresponds to locating a variable which appears in one equation and in no others. If we *wish* to calculate that variable rather than select it to be a decision variable, we *may* assign it to the equation. Because the variable does not appear elsewhere in the equations, we know that this calculation can be last among those which remain, as by this time all other variables in the equation will have been calculated or selected as decision variables.)

Assign as many of these variables as possible to be calculated. As they are assigned, note that they should be the last calculated among those not yet assigned. Decrement the remaining number of equations for each unit where such assignments are made. If the number of equations for any unit reduces to zero, delete that unit. Repeat this rule as often as possible.

(Decrementing the number of remaining equations for a unit by one for each variable so assigned is equivalent to deleting an assigned output variable and its equation from an incidence matrix as we did earlier. Deleting a unit whose number of equations reduces to zero simply means all of the equations are now deleted, so the unit is effectively deleted.)

In our problem all remaining unit parameters and all input and output stream parameters, which are for the flowsheet as a whole and which are unassigned, satisfy this rule. Figure 7.11 shows these assigned as outputs and the equation totals reduced appropriately. For the mixer, two input stream parameters are assigned to be calculated, and the number of equations remaining is reduced from five to three. The two reactor unit parameters as yet unassigned are made outputs and the equation count reduced by two, and so forth.

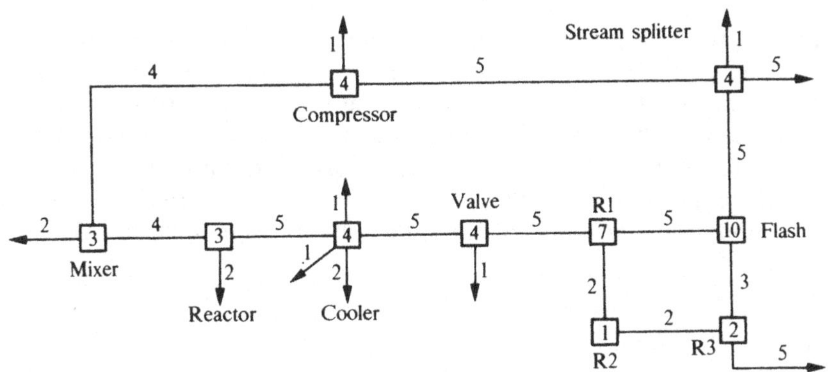

Figure 7.11. Rule B applied.

At this point rule B can be applied no further. The flowsheet must have a recycle computation in it at the outer level of computation. (It may well have recycle calculations within those for the units too, but our analysis here is only attempting to eliminate the recycle loops external to the unit calculations.) Figure 7.12 gives the results so far achieved. We must now apply rule C.

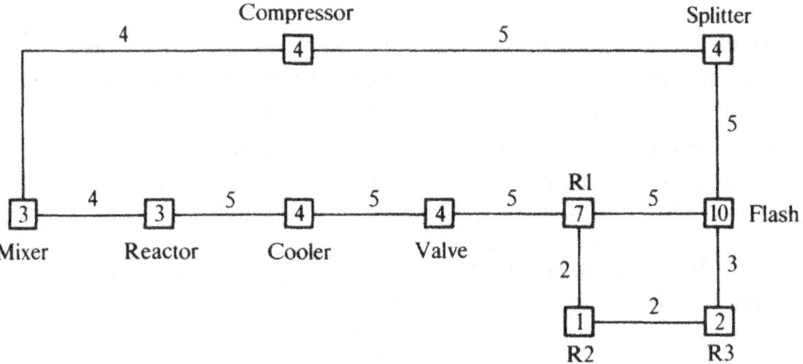

Figure 7.12. Results after rule B has been applied.

C. We could process a unit by rule A if we could create for that unit a situation where its unassigned variables become equal in number to the remaining number of equations it represents. This situation can be created if we arbitrarily assign some tear and/or decision variables. Here we find a block with the fewest required tear and/or decision variables and delete these variables entirely from the flowsheet. Then we apply rule A to the result. The deleted variables become tear and/or decision variables.

(Again, this rule is an extension of trying to create a single equation with but one variable left in it by declaring the rest to be 'known'. Being known means that the other variables must become decision variables or their values will have to be guessed, in which case they become tear variables. Rather than select which of the variables to calculate and which to declare as known, we can simply delete them all and the equation, knowing we shall make the actual assignment later.)

For the flowsheet, the mixer has eight associated variables and three equations, and thus would require five tear and/or decision variables. The reactor requires $9 - 3 = 6$, the cooler $10 - 4 = 6$, and so forth. Two of the ratio blocks (R2 and R3) and the flash require three tear and/or decision variables each. We choose the flash bottom product variables and delete three variables. Figure 7.13 shows these variables deleted as indicated by the dashed line.

Application of rule A (see figure 7.13) reduces the following units in the order listed to zero equations: flash, ratio block R1, bottom-right ratio block R3. The ratio block remaining, R2, has no remaining variables so its one equation can only be evaluated to see if it is zero.

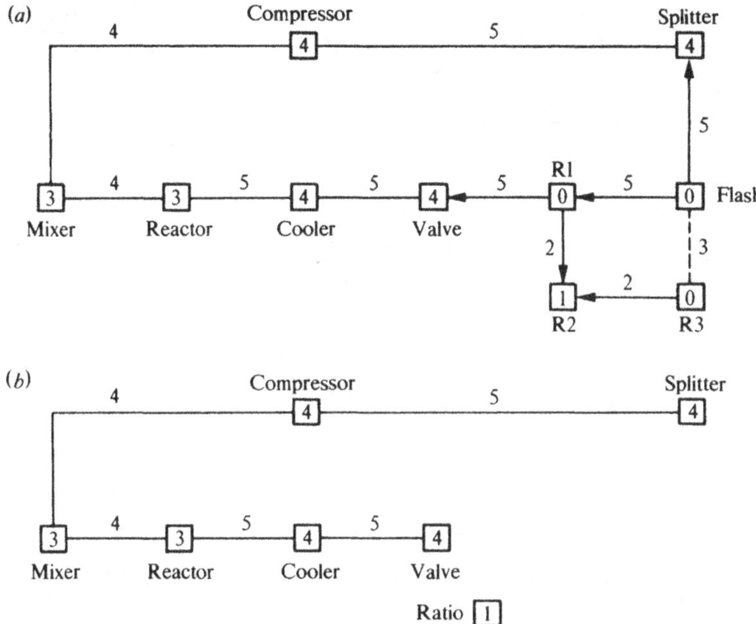

Figure 7.13. Application of rule C tears three variables and (using rule A) deletes flash and two of the ratio blocks (*a*), leaving block diagram (*b*).

(Again having a unit block remaining with some unassigned equations but no variables is equivalent to creating a single equation with all variables in it already known from previous calculations using other equations. That equation can then be used as an error function by rearranging it so it is of the form

$$g(x) = 0$$

and evaluating it to see if it is zero.)

Four more applications of rule C introduce one additional tear and/or decision variable each and are illustrated in Figure 7.14. We find we have introduced a total of seven tear and/or decision variables and have created four equations which have no associated variables; one is in the single equation ratio block and three are among the reactor equations. Our original analysis indicated that we had 16 degrees of freedom in the flowsheet, and 13 were taken up by design specifications – see figure 7.10. Three remain, therefore, and can be chosen from among the seven tear and/or decision variables we have found. At times they can be chosen to our advantage, permitting the iterations to be partitioned into two or more separate calculation loops.

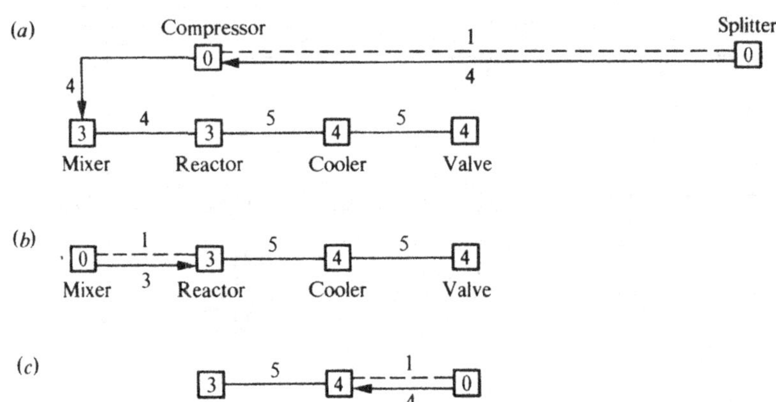

Figure 7.14. Four applications of rule C to fully reduce the block diagram.

The following synopsis of our discoveries so far aids us in finding if partitioning is possible. This step corresponds to our creation of the list in section 3.4.2.

1. Prechosen decisions
   - Feed stream — x_A, P
   - Reactor inlet — T
   - Reactor — Q
   - Cooling water in — T, P
   - Cooling water out — T
   - Flash bottoms — F, x_C
   - Flash — Q
   - Compressor — η
   - Compressor out — P
2. Tear/decision
   - Flash bottoms — 3 variables
3. Calculate
   - Flash feed (flash unit)
   - Top ratio block, R1
   - Bottom-right ratio block, R3
4. Evaluate
   - Bottom-left ratio block, R2 — 1 equation
5. Tear/decision
   - Compressor feed — 1 variable

6. Calculate
   Compressor feed (splitter)
   Compressor outlet (compressor)
7. Tear/decision
   Reactor feed   1 variable
8. Calculate
   Reactor feed (mixer)
9. Tear/decision
   Cooler outlet   1 variable
10. Calculate
    Cooler outlet (valve)
11. Tear/decision
    Cooler inlet   1 variable
12. Calculate
    Cooler inlet (cooler)
13. Evaluate
    Reactor   3 equations

Starting at the last step, we have three error functions which should be zero at the solution. We assign the last three tear/decision variables (items 7, 9 and 11) to this task and discover that no further error functions are included between the first of these tear/decision variables and the three error functions in step 13. So steps 8 to 13 are a loop of three tear variables and three error functions. The problem thus partitions at this point.

Step 4 gives us another function; we associate it with the 'closest' tear variable occurring before it. One of the three in the flash bottoms will suffice. The remaining three should become our decisions.

If we had already assigned both the tear/decision variables in steps 9 and 11 to be decision variables, the results would be quite different. Now the search for three tear variables for the error functions in step 11 would lead us all the way back to step 2. Before finding the third tear variable, we would have encountered the error function in step 5. It must then be a part of the same computation loop, and we would need to look for four tear variables in total. The whole calculation becomes a single computation loop as a consequence. Figure 7.15 illustrates the approach, comparing these two cases.

We are now left with checking whether a set of legitimite tear variables can be found as prescribed. This step is where we now examine the unit equations to see if we can meet the required information flow. We could choose arbitrarily the set: flash bottoms x_A, compressor inlet x_B, reactor inlet F and cooler outlet T. We need to ask whether for the splitter, for example, we can specify all five input stream variables (from the flash unit computation) and also the mole fraction x_B of B out into the compressor feed. This unit calculation is

	Case 1			Case 2	
	Number	Description		Number	Description
Tears/decisions	3 ← 0	1 tear, 2 decisions		3 ← 0	2 tears, 1 decision
Evaluate	1 — 1 loop			1 → 2	
Tear/decision	1	1 decision		1 ← 1	1 tear
Tear/decision	1 ← 0	1 tear		1 ← 2	1 tear
Tear decision	1 ← 1 loop	1 tear		1	1 preselected decision
Tear/decision	1 ← 2	1 tear		1	1 preselected decision
Evaluate	3 — 3			3 — 3	

Figure 7.15. Finding the best choice of decision variables to minimize the size of iteration loops. Starting from the bottom, the numbers along the arrows indicate the number of tear variables yet to be located before an independent calculation loop is defined.

1. Input from flash calculation
    Vapour stream F, T, P, x_A, x_B
    Input from routine guessing tear variable values
    Compressor inlet stream x_B
2. Calculate
    Compressor inlet stream F, T, P, x_A
    Bleed stream F, T, P, x_A, x_B
    Split ratio a

Clearly we cannot do this calculation, as x_B in the compressor inlet stream must equal the value for x_B into the splitter. A splitter only splits the total flow. The above calculation suggests these two x_B values are to be set independently. The only tear variable possible here is the flow F of the compressor inlet stream. Fortunately this choice can be made. We must examine all unit calculations now to choose the particular tears.

If no legitimate choice is possible, any algorithm for locating a solution procedure must backtrack and not allow the attempted information flow. Clearly the algorithm gets complicated, but it can be formulated. To get more complex than we have here is not within the scope of this book, so we shall stop at this point.

Although we proceeded in this analysis as though we were going to solve using tearing, we could use other methods. If we choose to use successive linearization, then our effort has identified four smaller *groups* of units which we can solve separately. The equations implied by steps 2 to 4 followed by the splitter and then the compressor and then followed by those in steps 7 to 13 can be solved as four groups of equations and not all together as a single group. Again we appeal to intuition that this partitioning is advantageous.

In chapter 6 describing the sequential modular approach to flowsheeting, we examined this same problem and had to solve all the units together and had nine tears at the flowsheet level. Thus we have reduced the number of tears to four and broken the problem into smaller parts by this treatment.

Earlier we mentioned that an incidence matrix formulation would be less tedious for a large block diagram. We shall present an example of such an incidence matrix and indicate briefly its use. Each basic block representation can be considered to be an equation set in the associated variables. Figure 7.16 gives an incidence matrix representation to these equations and variables for the block diagram in figure 7.2(*a*). We can analyse it much as we did in section 3.4.2 for solving nonsquare sets of nonlinear equations using tearing methods. Rather than eliminating incidences, we decrement the incidence counts and eliminate only those finally decremented to zero. Similarly we decrement the equation counts as they are used. Figure 7.17 illustrates its use.

	S1	S2	S3	S4	S5	P2	P3
Mixer (7)	7	7			7		
Reactor (7)		7	7			3	
Flash (14)			7	7	7		1

Figure 7.16. Incidence matrix for block diagram in figure 7.2(*a*). The number of equations for each unit is shown in parentheses. S, streams: S1, feed; S2, reactor inlet; S3, flash inlet; S4, flash outlet; S5, recycle. P, equipment parameters; P2, reactor; P3, flash.

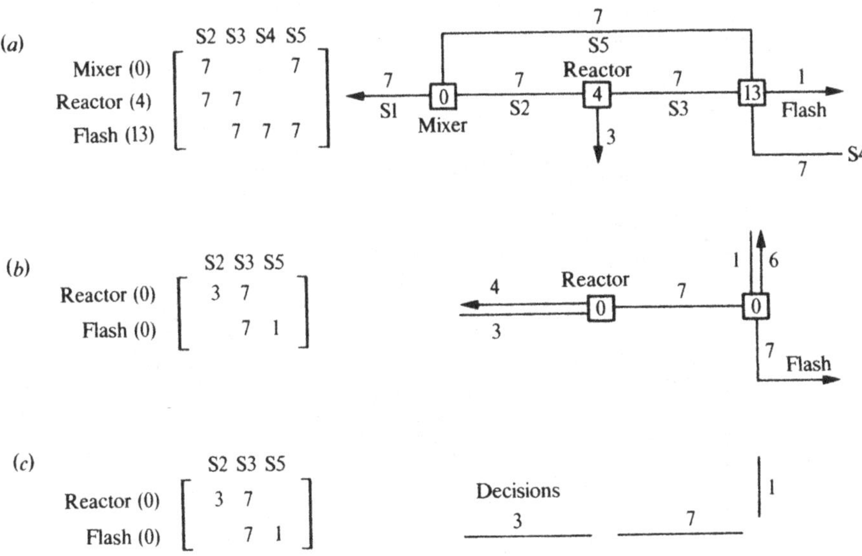

Figure 7.17. Use of an incidence matrix representation to develop a solution procedure.

7.4 Handling the iterated variables

The hierarchical approach to the solving of a flowsheet as suggested in the previous sections does imply that each unit will be solved completely when it is to be solved at the flowsheet level. If, however, the unit routine is automatically written so that the identity of its tear variables *and the values for its tear functions* are made known to the flowsheet routine, the functions can be converged outside the unit's equation-solving module. Figures 7.18 and 7.19 illustrate this difference. In figure

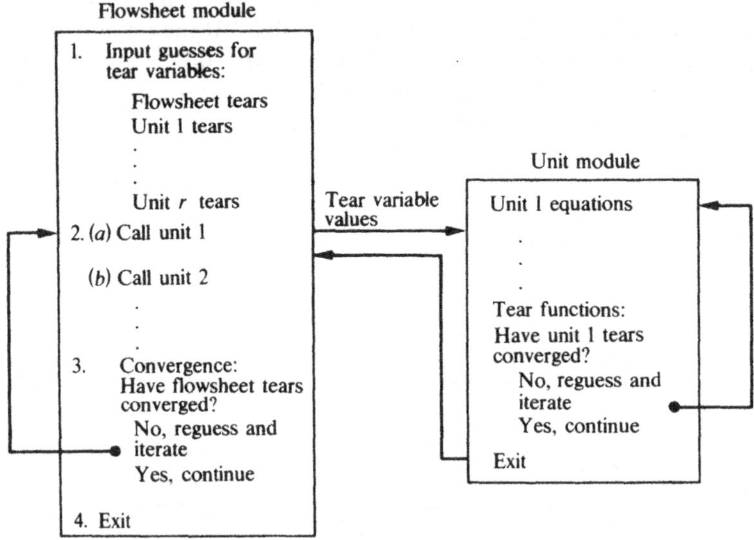

Figure 7.18. Converging tear variables within unit modules.

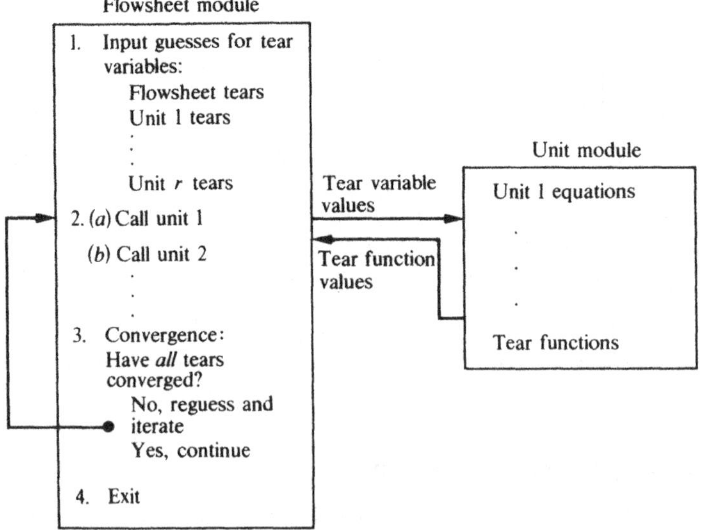

Figure 7.19. Not converging tear variables within unit module.

7.18, each unit module converges its tear variables each time it is entered. In figure 7.19, the tear functions are returned to the calling flowsheet module and are converged with the flowsheet tear variables. The above idea is based on work by Johns (1970).

A crude analysis can be used to guess at the relative merits of each approach. Assume a full multidimensional secant approach is to be used, and the time taken to solve n functions is proportional to $t^a n$ where t is the number of tear variables and a is an exponent conservatively valued at 3.

Suppose we have two units (unit 1 and unit 2); suppose also that unit 1, unit 2 and the flowsheet each have n_1, n_2 and n_f defining equations with t_1, t_2 and t_f tears respectively. Then the method implied in figure 7.18 would take a time θ_1 proportional to

$$(t_f^a n_f)\,(t_1^a n_1 + t_2^a n_2)$$

and the method implied in figure 7.19 a time θ_2 proportional to

$$(t_1 + t_2 + t_f)^a (n_1 + n_2 + n_f)$$

Table 7.1 gives some time ratios predicted by this approach for various values of n_1, n_2, n_f, t_1, t_2 and t_f for $a = 3$. It is only when the flowsheet tears are few, here no more than one, that the loops-within-loops approach implied by figure 7.18 is better – generally it is much worse. This analysis is crude and should only be used to get a qualitative feel for the problem.

Table 7.1. *Various time ratios for sample values of* n_1, n_2, n_f, t_1, t_2 *and* t_f

n_1	t_1	n_2	t_2	n_f	t_f	θ_1/θ_2 $\begin{cases} >1 \text{ One-level tearing better} \\ <1 \text{ Two-level tearing better} \end{cases}$
10	2	10	2	10	2	$12\ 800/6480 = 1.98$
100	2	100	2	100	2	$1.28\times10^6/6.48\times10^4 = 19.75$
100	2	100	2	10	2	$1.28\times10^5/4.536\times10^4\ /\ 2.82$
10	4	10	4	10	4	$8.192\times10^5/5.184\times10^4 = 15.8$
10	4	10	4	10	3	$3.456\times10^5/3.993\times10^4 = 8.66$
10	4	10	4	10	2	$1.024\times10^5/3\times10^4 = 3.41$
10	4	10	4	10	1	$1.28\times10^4/2.187\times10^4 = 0.585$
100	10	100	10	10	2	$1.6\times10^7/2.2361\times10^6 = 7.15$
1000	100	1000	100	10	2	$1.6\times10^{11}/1.66\times10^{10} = 9.66$
1000	100	1000	100	10	1	$2\times10^{10}/1.63\times10^{10} = 1.22$
1000	4	1000	4	10	1	$1.28\times10^6/1.46\times10^6 = 0.874$
10	2	10	2	100	10	$1.6\times10^7/3.29\times10^5 = 48.6$

7.5 Discussion

At the present time it is difficult to discuss any general trends evident for full-scale flowsheeting using equation-solving methods which are based on tearing, since these types of systems have not matured. We might conjecture the possibilities instead, and hopefully come close to considering what will exist.

In sections 2.3.4 and 7.3, we have already discussed the general structure of this type of flowsheeting system. The input is similar in content to that for any flowsheeting system except, and it is a very important 'except', that the user can readily add quite general specifications to the problem. Since the solution procedure is produced after those specifications are known, they can be incorporated and oftentimes used to advantage. They generally do not lead to additional and costly outer iterations as they do in the sequential modular approach.

The difficulties are obvious enough. The system has to develop the solution procedure automatically and to develop it so it will seldom fail. Otherwise the user will stop using the system.

Some aspects of this approach are much easier for the user. The writing of a user module to model a unit not already in the system is potentially much easier. The user need not concern himself with the solution procedure but rather only with the defining equations and constraints. (The constraints are to preclude flows etc. becoming negative during solving in case such a value leads to an attempt to take, for example, the logarithm of a negative number, or to division by zero.)

The equation-solving approach trades space for speed, and in the future systems it may also be trading space for a better chance to reach a solution. For the sequential modular approach, all units modelled by the same functional module require only one copy of the routine for that module. If three columns use a routine called COL3, only one copy of COL3 has to be present, and it is simply called with different data to distinguish among the columns. On the other hand, in the equation-solving approach each column may use the equations in COL3 but each may have a unique way of using them because of the different degrees of freedom selected for each, and because of the numerical considerations which may have been used to derive the solution procedure. Thus the equivalent of the routine for COL3 may appear three times in this approach.

For physical properties the potential for multiple copies could, for very large problems, preclude the idea of including these calculations as a set of equations rather than as a separately handwritten set of service routines. If computation time spent in the physical properties package for current systems is a measure of their relative number in a flowsheeting system, they could be 80% of the equation set. Thus a column with 1000 non-physical-property equations could give rise to an extra 4000 physical property equations.

The fact that the solution procedure can be tailored to meet the situation may, however, give two results: (1) a much faster execution time and (2) a better chance for converging to a solution. A highly integrated process with numerous interconnections among the units has been in the past extremely difficult to solve using the sequential modular approach.

The last comment deals with a real prediction, and that is that the systems will evolve using a hierarchical approach. Simon (1969), in a series of essays on the science of the artificial, gives a persuasive argument that complex systems are much more likely to evolve, given that they organize on a hierarchy, and one can see merit to this idea here. Thus the ideas of section 7.3 may underlie such future systems.

8

Simulation by linear methods

8.1 Introduction to linear simulation

Linear simulation, in the strict sense, is possible only when the performance and other characteristics of all the units in a process are known beforehand. In spite of this restriction, the method is a useful technique, particularly for preliminary simulations, or in other situations where reasonable guesses may be made as to the performance of units, but little precise information is available. The motivation for the study of this method is very simple, and can be seen by considering the recycle of a single component, as in figure 8.1.

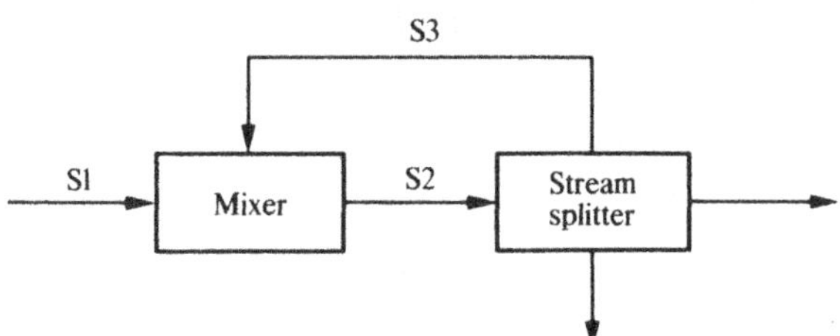

Figure 8.1. A simple recycle.

Suppose a fraction a of the material in stream S2 is recycled via stream S3, then the following equations can be written, for F_i, the flow in stream i.

$$
\begin{aligned}
\text{Mixer:} && F_1 - F_2 + F_3 &= 0 \\
\text{Stream splitter:} && -aF_2 + F_3 &= 0 \\
\text{Feed specification:} && F_1 &= F
\end{aligned}
\right\} \qquad (8.1)
$$

whence $F_2 = F/(1-a)$, $F_3 = aF/(1-a)$, by solution of the linear equations. By contrast, any process of guessing F_2 or F_3 leads to an iterative process, for which convergence will depend on the value of a.

Equations (8.1) contain, even in this simple example, three important types of relationship which arise in linear simulation. The first equation is an example of a conservation equation, the second of a constitutive equation, and the third an example of a constraint equation.

Conservation equations. These are equations describing the conservation of mass (or energy or momentum where appropriate), e.g. at the junction of flows of matter, or stoichiometric equations describing chemical reaction.

Constitutive equations. These describe the performance of some unit being simulated, e.g. the fraction of a flow recycled from a particular point, or the extent of a chemical reaction, or the performance of a distillation column.

Constraint equations. These describe particular constraints imposed upon the system that is being modelled, e.g. the composition or flow of some particular stream, which may be an input or an output stream, or indeed any other stream.

To achieve a complete solution it is necessary to find as many equations as there are variables in the system being studied. The systematic representation of a large problem with many recycles and other complicated interrelationships between variables by means of a linear simulation will often give insight into the essential degrees of freedom of the plant, which may well assist more detailed simulation.

The method has an important degree of flexibility. In the equations (8.1), the constraint equation could equally be replaced by an equation of the form

or
$$\left. \begin{aligned} F_2 &= G \\ F_3 &= H \end{aligned} \right\} \tag{8.2}$$

and either of these constraints, when used with the first two of equations (8.1) will obviously lead to satisfactory solutions. In subsequent sections this approach will be illustrated by its application to staged operations and to a management problem, and by a commercially available material balancing program, SYMBOL (CADC, 1973*b*).

8.2 Application to staged operations

The application to staged operations of the linear simulation method is not of great practical importance because graphical methods and variants of the Kremser equation are so readily applicable in the linear situation. The study of this area is of importance, however, as an introduction to general linear plant simulation and, given an easily available routine for solving sparse linear equations such as described in section 3.2.3, it may be that the equation-oriented approach described here would be a useful one. Two examples will be described in some detail.

8.2.1 Recovery of a volatile component in a special-purpose plant

The plant is shown diagrammatically in figure 8.2. It consists of three interconnected flash drums, each preceded by a heater in which the heat input is controlled so that one-third of the entering stream is vaporized. The vapour streams (S2, S5, S7) leave the flash drums in equilibrium

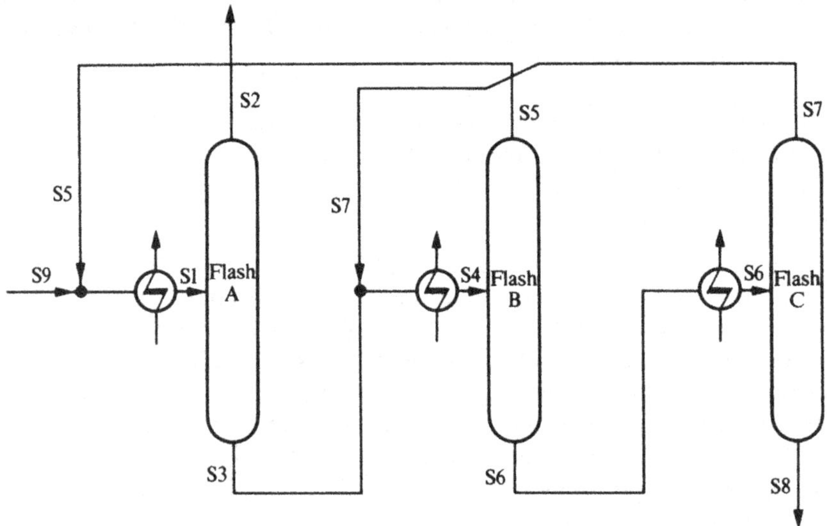

Figure 8.2. Three-flash-drum example.

with the liquid streams (S3, S6, S8). The temperatures and pressures on the three stages are so controlled by throttle valves and pumps (not shown) that the K-value for the more volatile component may be assumed constant at the value 4.2 on each stage. It is required to calculate the flows and relative compositions of the exit streams (S2, S8) and the recovery of the volatile component.

It is convenient to consider first the question of the total flows in each stream of this plant, and then later the flows of the more volatile component. There are, therefore, nine variables in this problem, and we shall discuss the ways of setting up nine linear equations from which the problem can be solved.

There are two simple mixing points in the plant, where streams S9 and S5 mix to give stream S1, and where streams S3 and S7 mix to give stream S4. For each of these points we can apply Kirchhoff's law that the sum of the flows must be zero. These equations are equations 1 and 2 of table 8.1. For flash drum A, we have that stream S2 is one-third of stream S1, which implies that stream S3 is two-thirds of stream S1, or that stream S3 is twice stream S2. These relationships are expressed by equations 3, 4, and 5 of table 8.1. A fourth relationship, equation 6, arises from the overall mass conservation of the flash drum. Not all these equations are independent, of course, but they are all valid equations for the flash drum and are listed to draw attention to the variety of information that is available. Similar groups of four equations 7, 8, 9, 10 and 11, 12, 13, 14 arise from flash drums B and C. There is also an overall mass balance from streams S9, S2 and S8. Finally, the scale of operations must be set. This could be done by setting any one of the nine streams to an arbitrary value, but to avoid unnecessary expansion of table 8.1, we have simply indicated the choice of F_9 to be unity. We have, therefore, a total of 24 valid equations from which to select a set of nine nonredundant equations to solve this problem.

Table 8.1 *Linear equations for example in figure 8.2*

Equation		
1	$F_1 - F_5 - F_9$	$= 0$
2	$F_4 - F_3 - F_7$	$= 0$
3	$F_2 - \frac{1}{3}F_1$	$= 0$
4	$F_3 - \frac{2}{3}F_1$	$= 0$
5	$F_3 - 2F_2$	$= 0$
6	$F_1 - F_2 - F_3$	$= 0$
7	$F_5 - \frac{1}{3}F_4$	$= 0$
8	$F_6 - \frac{2}{3}F_4$	$= 0$
9	$F_6 - 2F_5$	$= 0$
10	$F_4 - F_5 - F_6$	$= 0$
11	$F_7 - \frac{1}{3}F_6$	$= 0$
12	$F_8 - \frac{2}{3}F_6$	$= 0$
13	$F_8 - 2F_7$	$= 0$
14	$F_6 - F_7 - F_8$	$= 0$
15	$F_9 - F_2 - F_8$	$= 0$
16	F_9	$= 1$
17–24	Alternative constraints	

One such choice is the set (3, 4, 1, 7, 8, 11, 16, 12, 2) which is solved in section 3.2.3, and is set out in that order in table 3.7. This choice is made by considering the process unit by unit, describing the two mixers by equations 1 and 2, and each flash drum by the first pair of equations relating to that operation, and then applying one external constraint, 16, to the whole operation.

The setting up of this problem has been discussed at length because it illustrates one of the principal difficulties which face an engineer who attempts a linear simulation of a large chemical plant from first principles. Which of the many available equations constitute an independent set? In section 8.4 is described a system which allows the engineer to describe his plant within a clearly defined framework which takes the problem out of his hands, the program itself deciding which equations to set up.

The solution for the total flows for this problem is given in table 8.2.

Turning to the flows of the more volatile component, exactly the same set of equations may be set up, except that the value of the coefficients, 1/3, 2/3 and 2 will be different. Using the facts that (*a*) the vapour fraction of the inlet is $0.333 \times$ (total flow) and that the vapour composition is 4.2 times the liquid composition on the stage, it is readily shown that the vapour flow of the volatile component is 0.677 times the volatile component in the inlet stream. The same equations arise as shown in table 8.2 but with the numbers 0.677, 0.323, 0.477 replacing the numbers 1/3, 2/3, and 2 respectively. The actual composition of the feed stream S9 is not stated, so it is only possible to compute relative compositions. This can be done by leaving equation 16 unchanged. The solution to this set of equations follows exactly the same lines, and the relative flows of the volatile material are also set out in table 8.2. From these results it is easy to show that the recovery of the volatile matter in the overhead stream is 94%, and the relative concentrations in the overhead stream and the bottom stream are 2.02 and 0.11 times the input stream concentration respectively.

Table 8.2 *Results for figure 8.2.*

Stream	Total flows*	Volatile component flow*
S1	1.400	1.388
S2	0.466	0.940
S3	0.934	0.448
S4	1.200	0.574
S5	0.400	0.389
S6	0.800	0.185
S7	0.267	0.125
S8	0.533	0.060
S9	1.000	1.000

* Relative to input flow of unity

In this example, largely for convenience of description, two separate sets of equations have been set up, one for the total flows, and one for the more volatile component. In the more general situation, all the variables will be related in a single large set of equations, and no attempt will be made to partition the problem either by component or by groups of equations. It will be assumed that efficient sparse-equation-handling techniques similar to those described in section 3.2.4, and large enough to handle the largest problems likely to be encountered, are available.

8.2.2 Dilute gas absorption in a staged column

This example will be discussed more briefly and in more general terms than the last. Absorption of impurities in a gas stream by means of an inert oil which is effectively involatile is envisaged. The impurities are supposed present in small amounts, so that the flows of gas and oil are effectively unchanged on the column. The column is supposed isothermal and isobaric, so that uniform K-values for the impurity components may be assumed throughout the column. The nomenclature, stage numbering, etc. may be seen from figure 8.3.

For each impurity component i the following equations may be written in terms of the compositions (mole fractions) x and y in the liquid and gas phases respectively.

$$\left.\begin{aligned} Gy_{i,n+1}-Gy_{i,n}+Lx_{i,n-1}-Lx_{i,n} &= 0 \\ y_{i,n}-K_i x_{i,n} &= 0 \end{aligned}\right\} \tag{8.3}$$

$$i = 1, 2, \ldots, c \quad n = 1, 2, \ldots, N$$

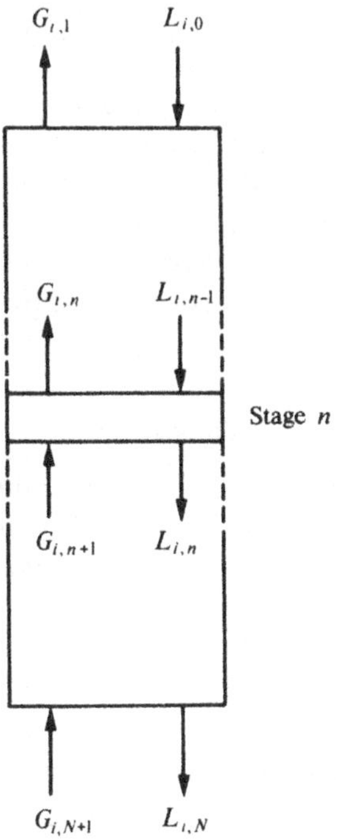

Figure 8.3. Gas absorption in a staged column.

Alternatively, the equations may be written in terms of the actual flows of components

$$G_{i,n+1}-G_{i,n}+L_{i,n-1}-L_{i,n} = 0 \Big\}$$
$$G_{i,n}-(K_iG/L)L_{i,n} = 0 \Big\} \qquad (8.4)$$

Since the example of section 8.2.1 used component flows, in this example we use the formulation of equations (8.3).

There are therefore $2c$ variables for each stage of the column, with, in addition, the $2c$ variables required to define the compositions of the inlet gas and liquid respectively. To solve for these we have c conservation relations for each stage, and c constitutive relations (equilibrium) for each stage, so that it is necessary to apply $2c$ constraints to achieve a solution. One obvious choice, case A, is to define the inlet gas and liquid compositions; this gives rise to a pure performance-type calculation, of which the results will indicate what can be achieved with a given number of stages. Another choice, case B, would be to set the outlet gas compositions to maximum permitted values on environmental considerations, and to set the liquid inlet compositions to achievable values; the results will then indicate the maximum concentrations in the inlet gas which can be permitted with the given number of stages and gas and liquid flowrates. Other combinations are possible. Table 8.3 sets out the equations in matrix form for case B, for two impurities and three stages.

It will be seen that the fundamental strategy is to set out the equations in any convenient order and to rely on sparse-matrix-solution techniques such as those described in section 3.2.3 to be able to solve the equations so set up. In particular, all the relevant variables are treated in a single large set of linear equations, whatever the structure or organization of the problem concerned. Notice also that the solution of all such properly posed linear problems is obtained by the once-for-all solution of the linear equations.

8.3 Application to a management problem

In this section we consider the application of linear simulation to a more managed and less technological problem. It will be seen that the method allows a solution to several important problems which are not suited for simulation by more detailed methods. It will also be seen how an approach of this type can allow insights into the overall operation of a process to develop naturally as the simulation is attempted.

8.3.1 Fundamental assumptions of the method

In simulation by linear methods as considered here, as opposed to the considerations in section 8.4, it is taken that the user is not concerned

Table 8.3 *Matrix equations for dilute gas absorption*

	$x_{1,0}$	$x_{2,0}$	$x_{1,1}$	$x_{2,1}$	$x_{1,2}$	$x_{2,2}$	$x_{1,3}$	$x_{2,3}$	$y_{1,1}$	$y_{2,1}$	$y_{1,2}$	$y_{2,2}$	$y_{1,3}$	$y_{2,3}$	$y_{1,4}$	$y_{2,4}$	Input
Absorption oil	1																p
compositions		1															q
Maximum gas compositions									1								r
on exit										1							s
Stage 1 mass balance	L		$-L$						$-G$		G						
		L		$-L$						$-G$		G					
Stage 1 equilibrium			$-K_1$						1								
				$-K_2$						1							
Stage 2 mass balance			L		$-L$						$-G$		G				
				L		$-L$						$-G$		G			
Stage 2 equilibrium					$-K_1$						1						
						$-K_2$						1					
Stage 3 mass balance					L		$-L$						$-G$		G		
						L		$-L$						$-G$		G	
Stage 3 equilibrium							$-K_1$						1				
								$-K_2$						1			

L, G: liquid and gas flowrates; K_1, K_2: equilibrium constants for components; p, q: inlet liquid compositions; r, s: outlet gas compositions with the detailed composition of a stream or with the detailed operation of a unit. It is assumed that a stream can be kept within specification by minor modifications to the operation of a unit, without altering significantly the performance of that unit. If this is assumed to be so, then plant performance records, if sufficiently detailed, will give important information which will permit the effect of small modifications of the process to be assessed, on the assumption that the individual units can be so controlled as to yield essentially the same performance. If plant performance records are not available, then design specifications may serve the same purpose.

Regarded as design or performance calculations, such calculations will necessarily appear crude, but it should be emphasized that the calculations themselves are not approximate. The linear equations are themselves solved exactly, and the results achieved are an exact solution to the problem as posed. The crudity, if any, is in the setting up of the linear equations: that is to say, in the assumption that each operation can be expected to yield similar results under the perturbed conditions as under the original conditions.

For example, consider the operation of a reactor and separation train subplant in this context, and assume that, from plant records, 80% of the incoming material is converted to the desired product, 3% to waste products, and that 15% is recycled to the reactor. For the purposes of linear simulation it may be assumed that this same performance may be achieved by suitable adjustments of reactor temperature, pressure, etc. and by suitable changes of operation within the separation train, whatever changes, within limits, are made to the incoming materials. If the changes are large, then the assumption becomes less plausible, and the numerical factors involved may have to be re-assessed.

We assume, therefore, that sufficient information is available to relate the main products, utilities, and materials consumption of any plant unit by linear relationships. In a simple distillation column, for example, we assume that the distillate rate is kept at a known fraction of the feed rate, and that the steam to the reboiler and possibly the cooling water requirement is linearly related to the distillate rate. Linear equations which define all of the quantities in terms of any one of them can be set up. Some quantities may be known to be at the manager's disposal – the division of material between two parts of the process, for example. Such quantities will be represented either by assigned values which can be varied at will, or by degrees of freedom which can be taken up by requiring, for example, a certain balance between two of the products from the plant. It is possible to assign the amounts of the main flows into the plant, but often more useful to replace such assignments by an equivalent number of demands for products.

The inclusion of nonprocess materials such as steam and cooling water has been mentioned, but such nonmaterial items as cost, power requirements, and even manpower may be included provided they can be linearly related to the main process variables. In the next section we show the application of the method to a simplified oil refinery scheme, and discuss in more detail the questions which can be asked and answered.

8.3.2 Application to an oil refinery

The model oil refinery which we will consider is shown in figure 8.4; it is a considerable simplification of the original plant to which the method was applied using the problem description language described in the next section. Only the process streams are shown in the diagram, but in addition cooling water, stream consumption, electric power consumption and fuel gas consumption may be taken into account where appropriate figures are known for the main process units relating such utility consumption to the flows in the process streams.

From plant records for a particular crude oil, performance factors were established which described the behaviour of the plant under steady running conditions. For the crude distiller, for example, the amounts in each of streams S1, S2, S4 and S5 were related to the amount of stream S3. At the same time the amount of fuel gas required was related to the amount of crude oil, stream S1, and the amount of steam required for side strippers to the sum of streams S3 and S4. Similar relationships were established for all the other operations, the catalytic cracker being represented by the fractions of stream S12 which went to streams S14, S20 and S21 respectively, together with the amount of steam generated by the process per unit of stream S12 treated. Gasoline blending was treated as a simple addition of the streams concerned.

Each unit process within the refinery can then be regarded as defined completely in terms of its inputs. We can then turn to consider the refinery as a whole. Suppose we start with a known flow of crude oil to the crude distiller. It is clear that the flows of all the output streams from the distiller are known. By following each of the streams in a feed-forward direction it becomes clear that every stream in the refinery can be calculated if one other piece of information is known. The missing information is the amount and direction of the cross-flow in stream S8 between stream S7 and stream S12, or between stream S4 and stream S9. Once this is known it is a simple matter to solve the whole refinery sequentially in a feed-forward manner.

It should be noticed that this is true whether or not the exact numerical relationships describing the performance of the units are known. Thinking in terms of linear relationships has merely helped to

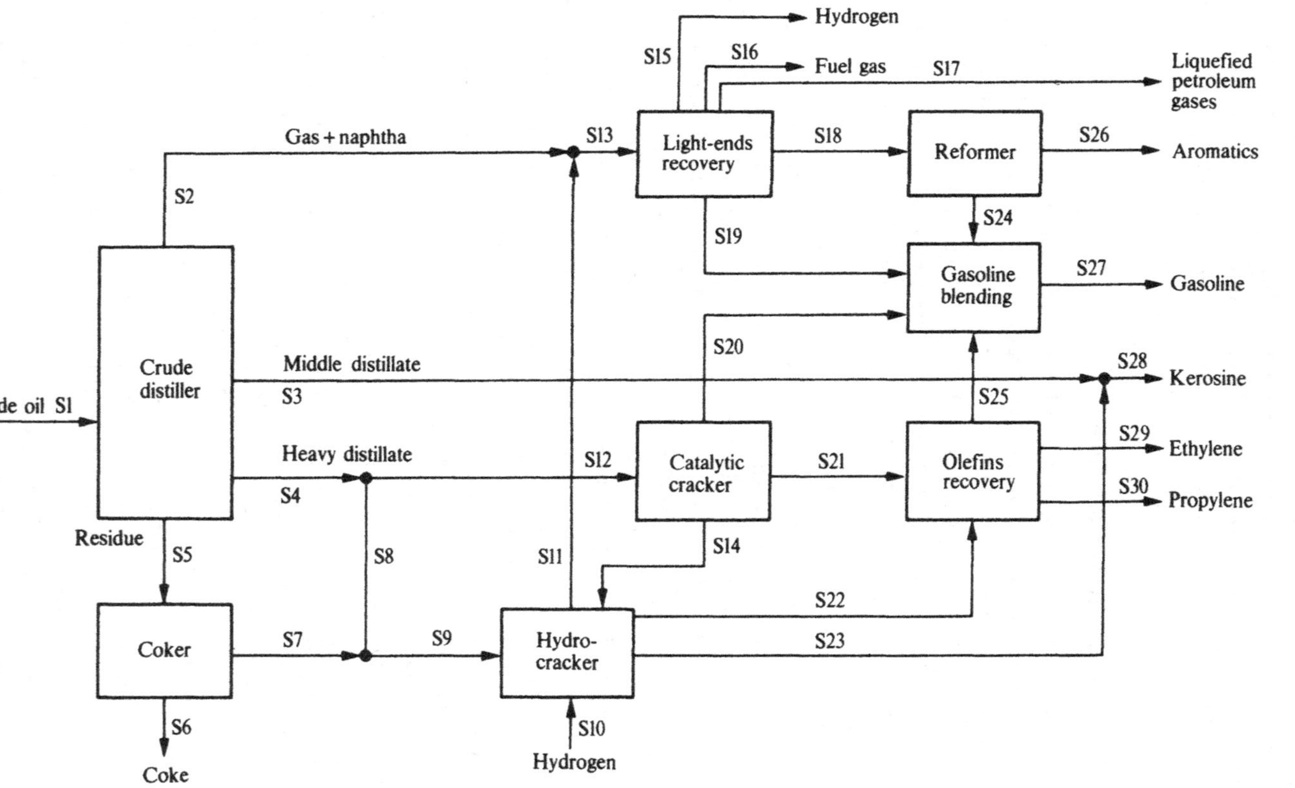

Figure 8.4. Simplified oil refinery scheme.

identify the main operating control variables of this plant, which we have identified as the scale of operation (the flow of crude oil) and the relative amounts of the feed to the catalytic cracker unit and the hydro-cracker (the cross-flow in stream S8).

However, these prove not to be the most useful constraints from the point of view of the manager operating this refinery. His main interests are, rather, how to meet the seasonal variation in demand for his two main products, gasoline and kerosine, and what would be the effect of meeting these demands on the production of by-products and the consumption of crude oil. He has a subsidiary interest in the costs, and utility consumptions, under these seasonal variations. Having identified that there are two constraints in this situation, it is clear that one constraint can very readily replace another in the set of linear equations representing the refinery. It is easy enough to set demands for gasoline and kerosine and to solve linear equations to give answers to all the flows in the refinery. These flows can then be checked against known limitations of throughput for the various unit processors.

However, other operating constraints can be applied, if desirable, to achieve solutions applicable in other situations. For example, if it is known that the limiting capacity of the plant is the feed to the catalytic cracker, stream S21, then this can be used as one of the constraints, leaving considerable choice of second constraint (e.g. to maintain the seasonal ratio of gasoline to kerosine). Another interesting possibility arises from the fact that usually there is plenty of hydrogen available from other operations on the same site. If these are closed down, it may be necessary to run the refinery at, or near, hydrogen balance. Again, the effects of such a constraint can readily be seen by equating stream S10 to stream S15 as one of the constraints and then setting, say, the demand for gasoline as another constraint. Within the limitations of the linear modelling of the unit processes within the refinery, none of these combinations of constraint poses any more difficulty for solution of the equations than another. It is simply a matter of replacement of one linear constraint equation by another.

8.3.3 Setting up the linear equations

The setting up of the linear equations for a linear simulation of, for example, an oil refinery plant as described in the last section is not a trivial matter. One approach to this is to provide the user with a problem description language. This allows the user to describe the problem in a limited type of language in a rather similar way to that in which a programmer describes an algorithm in FORTRAN or ALGOL. The advantage of such a system is the ease with which a user can assemble and check the substantial amount of information that is required to simulate a moderate-sized chemical plant. The disadvantage is the computing time required for the interpretation of the language, which is usually very much longer than for the interpretation of numerical datasets.

Such a system may allow the use of alphameric strings, perhaps of limited length, to designate the stream flow quantities, related utilities, and costs, in a manner very similar to the use of simple variables in FORTRAN. Simple codewords may then describe the types of equation to be set up. For example, the crude distiller of the previous section might be represented by the codeword SPLIT, which sets up a mass balance for the named streams, and then sufficient ratios to define the overall behaviour of the column. A description in such a language, preceded by a comment, might then appear as

```
NOTE THE MAIN CRUDE DISTILLER MATERIAL FLOWS
SPLIT CRUDE = TOPS, MDDIST, HYDIST, RESID
                CRUDE/MDDIST  =  2.80
                HYDIST/MDDIST =  1.02
                RESID/CRUDE   =  0.13
```

Constraints, such as setting particular demands for gasoline and kerosine, may be set using the DEMAND codeword:

```
NOTE   SET DEMANDS FOR GASOLINE AND KEROSINE
DEMAND                GASOL = 16 500
DEMAND                KEROS = 22 050
```

Alternatively, if other constraints are to be used, such as setting stream S8 (figure 8.4) to a fixed ratio of the heavy distillate feed to the catalytic cracker, while maintaining a fixed gasoline production, one might have

```
NOTE   ALTERNATIVE CONSTRAINTS
DEMAND                GASOL = 16 500
RATIO                 STRM8/HYDIST = 0.18
```

In addition to the codewords already mentioned, such a language might have simple 'unit operations' such as MIX, FEED, REACT, INPUT, simple 'mathematical operations' such as ADD, SET, SUB-TRACT, RATIO, EQUATE, and the facility for incorporating a general linear equation relating any number of variables. The language might also contain instructions for printing the values of desired variables in groups, collating all the steam flows required and generated, for example, and also facilities for data modification and rerunning of problems.

Such a system, known as SYMMAN, has been implemented on a PDP–10 computer (Johnson, 1974).

8.4 The SYMBOL system for material balancing in linear systems

The system to be described has been used in its present or closely similar forms at Cambridge University since 1970 (Hutchison, 1974) and also more recently in other universities as a teaching aid, and in association with Design Projects. It is commercially available (CADC, 1973*b*). Other systems are known to exist, but this is believed to be the most fully developed such system, and offers very clear assistance to the user in the compilation of his input information.

The SYMBOL system is concerned only with material balancing, and only with linear equations. But the system is designed to hide from the user the fact that he is working with linear equations, by compelling the user to describe his process by means of a strictly limited set of unit operations. There are only five classes of operation, though each may be used in a number of modes. These are set out briefly in table 8.4. These five classes of operation may be used to simulate the unit operations of a real plant, though not necessarily in a one-to-one fashion. A cascade may represent a gas absorption column, for example, in simple one-to-one relationship. A real reactor system will normally have to be represented by one or more symbolic mixers, and one or more reactors corresponding to the number of independent chemical reactions taking place in the system. On the other hand, a train of crystallization tanks with associated filters may be represented by a simple separator, if the detailed flows within the train are not of interest.

It will be seen that modes 1 of source, separator and reactor are insufficient to define completely the exit flow from the source, separator or reactor (given the input flow to the separator or reactor). It is the presence of these incomplete definitions which gives the system its flexibility. This flexibility is shown by the need for, and existence of, the additional constraint modes shown at the end of table 8.4. Such additional constraints can be used, for example, to set the scale of operations to meet a desired end product production, or to meet purity constraints at one or more points in the process, or to represent physical limitations on the operation of the plant.

The collection of the data for such a system is made easier by arranging for it to be done in phases, using a computer to print or type data preparation forms which are specific to the problem in hand as the data preparation progresses. In this scheme there are three phases to running the SYMBOL system, which get progressively more complex as the final numerical simulation is approached. Input to the first data preparation program is simply a title for the simulation, and then the requirements of the whole process, i.e. the number of streams, components, sources, mixers, separators, cascades, and reactors. From this program is obtained a printed form on which the modes of the units may

Table 8.4. *Unit operations and modes in the SYMBOL system*

Operation	Mode	Information required
Source	1	Composition only of the source
	2	Composition and a flowrate
	3	Composition and ratio of a flowrate to a flowrate in any other stream
Mixer	—	Any number (implementation dependent) of streams may be mixed
Separator	1	Composition only of one of the exit streams
	2	As mode 1, with a flowrate of the exit stream
	3	As mode 1, with a ratio of a flowrate in the exit stream to a flowrate in any other stream
	4	Fractions of the flowrates of the input stream which go to a named output stream
	5	The fraction of all the flowrates of the input stream which goes to a named output stream. Simple flow divider
	6	As many independent ratios of flowrates of streams anywhere in the plant which may serve to define the performance of the separator. Flexible, but use with care
Cascade	1	Transfer in one direction between two phases. Fractions of the flowrates from one input phase which occur in a named output phase
	2	Transfer in two directions between two phases. Fractions of the flowrate from each input phase which occur in named output phases
Reactor	1	Stoichiometry only of the reaction
	2	Stoichiometry of the reaction. Fractional extent of the reaction based on a named reactant
	3	Stoichiometry of the reaction. Ratio of a flowrate in the exit stream to a flowrate elsewhere in the plant
Additional Constraint		
	1	Ratio of a flowrate in a stream to a flowrate in the same or another stream
	2	Value of a flowrate in a stream

A flowrate in a stream denotes the flowrate of a component or the total flowrate in a stream, and may be in mass or mole units provided either is used consistently throughout the simulation.

be specified, and the interconnections of the units specified by identifying which streams are inputs or outputs to particular units. When this form is completed, the information on it is used as input to a second data preparation program, which then prints a second form, specific to the problem, on which are provided spaces for all the numerical pieces of information necessary to define the plant. It is a particular function of this second data preparation program to indicate if, and how many, additional constraints are required. When this second data preparation form has been completed, the information on it is used as input information to the final, computation stage of the SYMBOL system. The overall information flow is seen schematically in figure 8.5, and examples of the printed forms will be discussed in the simple example of the next section.

8.5 A simple example of the use of SYMBOL

This simple example is introduced to illustrate some of the features of the SYMBOL system. The process concerned is the oxidation of butane

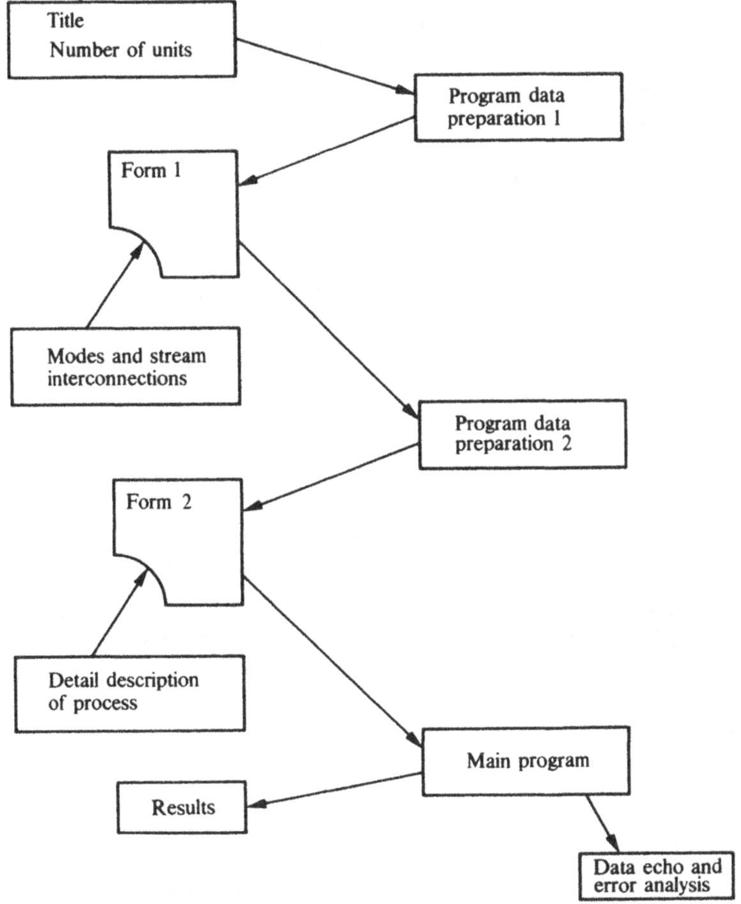

Figure 8.5. Information flow in the SYMBOL system.

using 95% oxygen–nitrogen mixture to produce principally acetic acid, with water and carbon dioxide as coproducts and a mixture of oxygen-containing compounds characterized as LOL (light organic liquids) for further purification. The oxidation takes place at about 150 °C and 10 bar, although this information plays no part in the computation. The block diagram of the process is shown in figure 8.6(*a*) and its representation in terms of SYMBOL units in figure 8.6(*b*).

Mass units rather than mole units are used throughout. For simulation purposes, seven components are assumed present throughout, butane, acetic acid, LOL, water, carbon dioxide, oxygen and nitrogen. In SYMBOL the total flow is always regarded as an additional component,

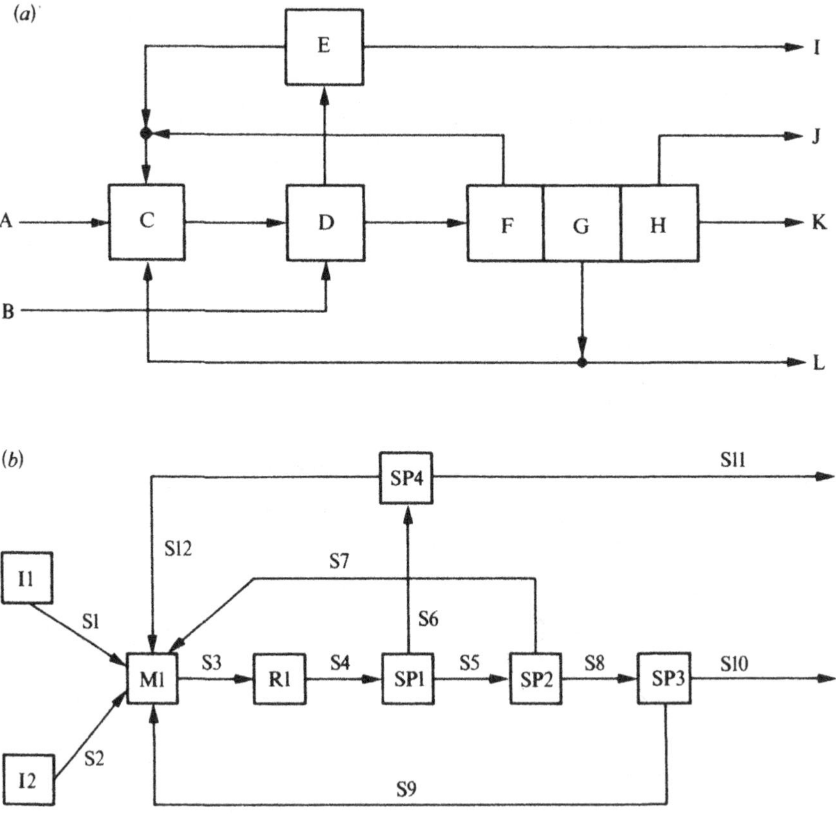

Figure 8.6. Simple example for using SYMBOL system. (*a*) A, butane feed; B, oxygen feed; C, butane–acid mixer; D, reactor; E, butane condenser; F, butane recovery; G, acid recovery; H, product separator; I, waste gases; J, product LOL (light organic liquids); K, waste water; L, product acid. (*b*) S1, butane feed; S2, oxygen feed; S3, reactor feed; S4, all reactor products; S5, liquid reactor product; S6, reactor take-off gases; S7, recycle butane; S8, products; S9, recycle acid; S10, products (LOL and acid); S11, waste gases, S12, recovered butane.

so that references may be made, in setting up the data, to component 8, which is the total flow. The plant must run under several constraints:

1. The butane must be dissolved in the acetic acid before it enters the reactor. Here we take that the weight of butane in the reactor feed must be 0.65 times the weight of acetic acid in that feed.
2. Experience indicates the 20% of the oxygen passes through the reactor unchanged.
3. The oxygen concentration in the reactor gas take-off must not exceed 6.5% of the total to avoid a potential explosion hazard.
4. The empirical stoichiometry of the reaction under these conditions is given in table 8.5 in mass units.

Table 8.5. *Empirical stoichiometry for butane oxidation*

Butane	−1.0 ⎫	Reactants
Oxygen	−1.69 ⎭	
Acetic acid	1.01 ⎫	
LOL	0.42 ⎬	Products
Water	0.58 ⎪	
Carbon dioxide	0.68 ⎭	
Nitrogen	0.	Inert

From the SYMBOL diagram, in which the streams are numbered, the input data to start the first stage of data preparation are the two lines

TITLE FOR BUTANE OXIDATION EXAMPLE
12   7   2   1   4   0   1

Use of the first program produces the form shown in figure 8.7 into which may be written the information necessary to specify the modes and interconnections of the units. The interconnections are fairly obvious, but the choice of modes for some of the units is not, and these will be discussed in some detail.

The first source is chosen to be mode 2 so as to set the scale of operation. The second source is mode 3, which by reference to table 8.4 requires, in addition to the composition, the ratio of two streams. This choice permits the second constraint to be met, by setting the flowrate of oxygen in this second source to be 5 times the flowrate of oxygen in stream S6. The first separator is really the disengagement space of the bubbling reactor. Mode 6 is chosen, which allows the specification of seven ratios to describe the separation. The first may be used to ensure that the explosive limit criterion is reached, the others describe the separation, complete or otherwise, between the two phases. Separators SP2 and SP4 are chosen to be mode 4 separators, whose specifications will each require seven split factors, while separator SP3 is chosen to be

```
SYMBOL                    FORM 1

TITLE FOR BUTANE OXIDATION EXAMPLE

STRMS  COMPS  SOURCS MIXERS SEPS  CSCDS  REACTORS
* 12 **   7 **      2 **     1 **     4 **    0 **       1 **


SOURCES
        MODE  STREAM  IDENTIFIER
  1   *  2  **  1 *
  2   *  3  **  2 *

MIXERS
        MODE  OUTPUT    INPUTS
  1   *  5  ** 3 **   1 **  2 **  7 **  9 **  12  **      **     *

SEPARATORS
        MODE  INPUT   OUTPUTS
  1   *  6  ** 4 **  5 **  6 *
  2   *  4  ** 5 **  7 **  8 *
  3   *  1  ** 8 **  9 ** 10 *
  4   *  4  ** 6 ** 11 ** 12 *

REACTORS
        MODE  INPUT   OUTPUT
  1   *  2  **  3 **  4 *

END OF FORM 1
```

Figure 8.7. SYMBOL system, form 1.

mode 1, requiring the composition only of one of the output streams. It is this choice which gives another degree of freedom permitting the operating constraints to be met. The mode chosen for the reactor is 2, which will permit the stoichiometric coefficients to meet the empirical coefficients, while leaving the extent of conversion as an important simulation parameter.

These choices (and others which could have been made) will enable a simulation which meets all of the constraints imposed on the system to be made.

When form 1 has been completed, the information on the form is sent as input to the second data preparation program, from which is obtained a second data preparation form, with spaces for each piece of information required to complete the simulation, figure 8.8. Completion of the second form, with the detailed figures necessary to describe the simulation, means that all the data has been assembled for a successful simulation of the problem. Note in particular that the final piece of information, the additional constraint, is used to control the butane–acetic acid ratio, although the degree of freedom which gives rise to the additional constraint is associated with the separator SP3. The mass balance which is finally obtained (for an extent of conversion of 0.12 based on butane) is shown in figure 8.9. It turns out that the extent of conversion is quite critical. A figure not much above 0.14 gives rise to negative flow quantities in some streams, indicating that under these conditions not all the constraints can be properly met.

8.6 A complex example of the use of SYMBOL

The SYMBOL system has been applied to several large chemical plants, of which one of the more interesting is the Bayer alumina preparation plant described by Crowe *et al.* (Crowe *et al.*, 1971). This process, as described by the SYMBOL diagram of figure 8.10, has 61 streams, which, with 10 components and the total flow for each stream, makes a total of 671 variables. It is also an extremely involved recycle process, and, moreover, it has to meet some important operating constraints. The main emphasis on the discussion in this section will be on how to set up the SYMBOL representation to meet the many constraints, and how changes to these constraints can provide answers to important operating questions.

In a situation of this sort, where many operating conditions have to be met, it is important for the engineer to have in hand as many degrees of freedom as possible. In this simulation it is possible, for example, to set all the sources to mode 1 (see table 8.4), thereby giving rise to 10 degrees of freedom. Then the crystallizers, which are represented by reactors R4, R5, and R6, and which are in fact operated to meet an analytical criterion on their output streams, may be set to mode 1.

Finally, separator SP15 which removes a small stream of material for a calcination process to remove impurities may also be set to mode 1, giving a total of 14 degrees of freedom, or additional constraints, to be satisfied before a solution can be achieved. The detail behaviour of all the units may be obtained from plant records or from detailed simulation by other methods.

One set of additional constraints which may be set up in order to achieve a simulation are as shown in table 8.6. Such constraints are indicated as C1, C2, etc., in figure 8.10. It will be noticed that these constraints cover a wide variety of types of operating conditions.

When a successful simulation has been achieved with this set of

SYMBOL   FORM 2
TITLE FOR BUTANE OXIDATION EXAMPLE

STRMS COMPS   SOURCS MIXERS SEPS CSCDS   REACTORS
* 12 ** 7 **   2 **   1 **   4**   0**   1 **

SOURCES
 1   *   2 **   1 **
      <   1.   >< 0.   >< 0.   >< 0.   >< 0.   >
      <   0.   >< 0.   >
      *   1 *< 100.   >
 2   *   3 **   2 **
      <   0.   >< 0.   >< 0.   >< 0.   >< 0.   >
      <   .95   >< .05 >
      *   6 **   4 ** 6   *<   5.0   >

MIXERS
 1   *   5 **   3 **   1 **   2 ** 7 **   9 **   12 **
SEPARATORS
 1   *   6 **   4 **   5 **   6 **
 1   *   6 **   6 **   6 **   8 *<   .065   >
 2   *   6 **   2 **   6 **   8 *<   0.   >
 3   *   6 **   3 **   6 **   8 *<   0.   >
 4   *   6 **   4 **   6 **   8 *<   0.   >
 5   *   6 **   5 **   4 **   5 *<   1.   >
 6   *   6 **   6 **   4 **   6 *<   1.   >
 7   *   6 **   7 **   4 **   7 *<   1.   >
 2   *   4 **   5 **   7 **   8 **
      *   7 *
      <   1.   >< 0.   >< 0.   >< 0.   >< 0.   >
      <   0.   >< 0.   >
 3   *   1 **   8 **   9 **   10 **
      *   9 *
      *   0.   >< 1.   >< 0.   >< 0.   >< 0.   >
           0.   >< 0.   >

```
4   *   4 **   6 **   11 **     12 **

    *   12 *

    <   .97    >< 1.          ><  1.  ><  1.  ><   0.   >

    <   0.     >< 0.          >

REACTORS

1   *   2 **   3 **    4 **

    *   1 *<  .12   >

    <   1.     >< −1.01        >< −.42 >< −.58 >< −.68 >

    <   1.69   >< 0.          >

EXTRA CONSTRAINTS

1   *   3 **   1 **    3 **    2 *<  .65   >

END OF FORM 2
```

Figure 8.8. SYMBOL system, form 2.

```
SYMBOL                          MASS BALANCE

TITLE FOR BUTANE OXIDATION EXAMPLE
```

STREAM	1	2	3	4	5
1	100.0000	0.	719.2654	632.9536	176.6819
2	0.	0.	1106.5622	1193.7371	1193.7371
3	0.	0.	0.	36.2510	36.2510
4	0.	0.	0.	50.0609	50.0609
5	0.	0.	0.	58.6921	0.
6	0.	182.3338	182.3338	36.4668	0.
7	0.	9.5965	9.5965	9.5965	0.
TOTAL	100.0000	191.9303	2017.7579	2017.7579	1456.7309

STREAM	6	7	8	9	10
1	456.2717	176.6819	0.	0.	0.
2	0.	0.	1193.7371	1106.5622	87.1750
3	0.	0.	36.2510	0.	36.2510
4	0.	0.	50.0609	0.	50.0609
5	58.6921	0.	0.	0.	0.
6	36.4668	0.	0.	0.	0.
7	9.5965	0.	0.	0.	0.
TOTAL	561.0270	176.6819	1280.0490	1106.5622	173.4868

STREAM	11	12
1	13.6882	442.5835
2	0.	0.
3	0.	0.
4	0.	0.
5	58.6921	0.
6	36.4668	0.
7	9.5965	0.
TOTAL	118.4435	442.5835

```
OUTPUT COMPLETE
```

Figure 8.9. Output from SYMBOL system for example problem.

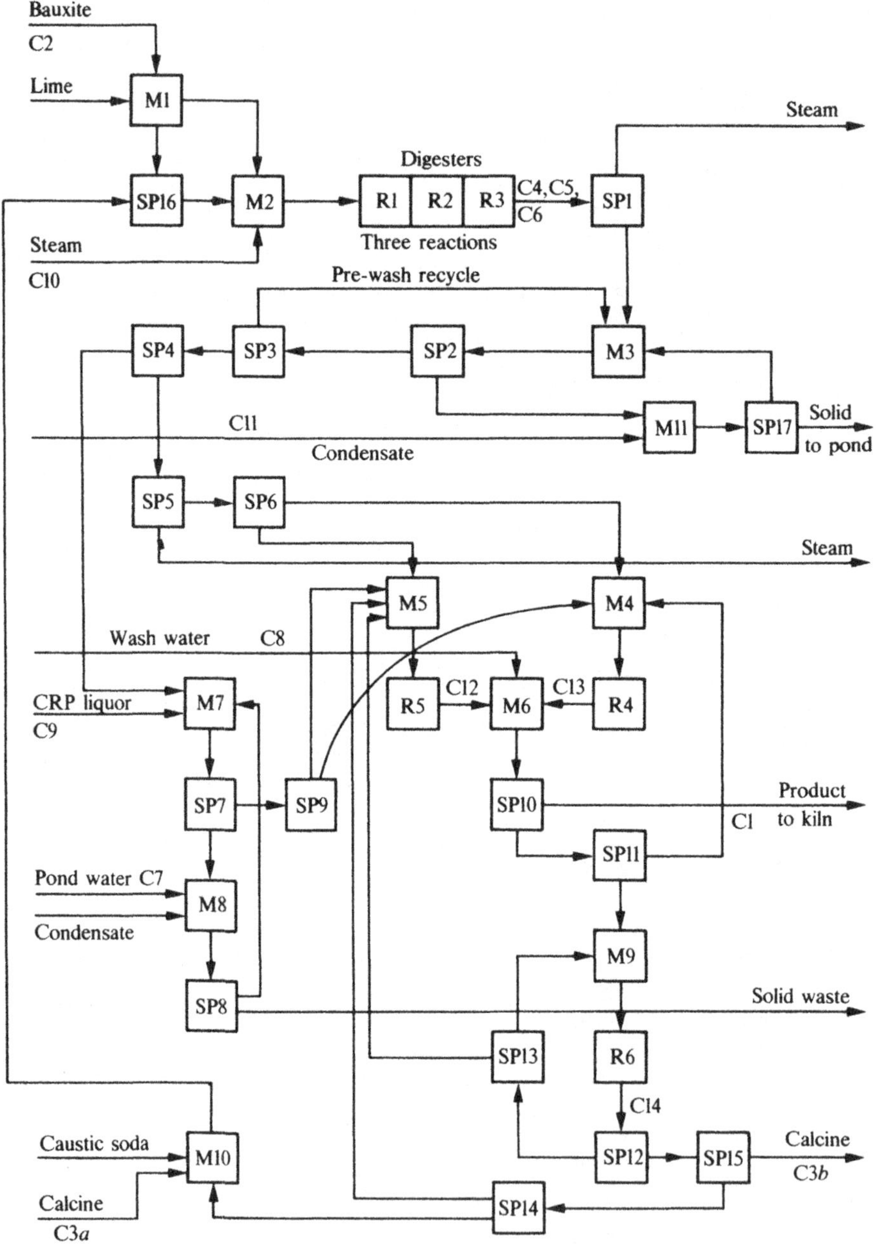

Figure 8.10. SYMBOL system for Bayer plant.

constraints, consideration may be given to some other possible constraints. The first and simplest is to set the flow of liquor (C9) to zero. Changes in the production rate, the pond water rate (C7) and the condensate rate (C11) may also be investigated successfully.

Table 8.6. *Additional constraints required for a simulation of a Bayer alumina plant (see figure 8.10)*

C1	Main production rate
C2	Lime-to-bauxite ratio
C3a,b	Calcine input rate related to calcine purge rate.
C4 C5 C6	Analytical requirements governing the composition of the exit stream from the digesters.
C7	Pond water limited by pump capacity
C8	Ratio this wash water to the main stream flow
C9	Ratio this liquor to total production
C10	Ratio steam into this preheater to the total flow. In reality this is a temperature control, but a ratio can be shown empirically true
C11	Ratio this condensate to the total production
C12 C13 C14	Ratio of aluminate to caustic soda at the exit of the three precipitators. This represents actual analytical control

It may readily be seen what a powerful technique is provided by linear simulation, particularly in its ability to meet arbitrary (by which is meant chemical or operational) restrictions on the process flow quantities or ratios at various points.

One difficulty with the method is that the operating characteristics of the various symbolic plant items have to be known beforehand, from plant records or by some other simulation. If they are correctly known for one set of the constraint parameters, they cannot be known exactly for a different set. Of course, an iterative approach is possible, with an overall simulation used alternatively with detailed simulation until a converged solution is reached, but this technique is really related to the quasi-linear methods of the next chapter.

One final point needs to be made about methods of linear simulation in general. However rough the approximation may be between one or more symbolic units and the plant item or items which they represent, the calculated material flows necessarily exhibit a material balance correct to within the rounding errors of the solution process. These flows therefore represent a feasible operating state of the chemical plant, satisfying all the imposed constraints, and obtainable, in principle, by some perturbation of the operating characteristics of the actual plant items.

9

Simulation by quasi-linear methods

9.1 Introduction to quasi-linear methods
9.1.1 Origination of the method

The essence of this method is very simply stated once simulation by linear methods, as described in the previous chapter, has been fully grasped. In quasi-linear methods the same fundamental approach is adopted, that is to say, a large set of equations consisting of conservation equations, constitutive equations and constraint equations is set up. Of these, the conservation equations are strictly linear already, while the constitutive equations are replaced by some linear approximation. Constraint equations are often linear in form, and these may be taken over unchanged, but if not, these too may be replaced by some linear approximation. The resulting set of linear equations is then solved to obtain a trial solution, and from this trial solution the linear approximations are updated to provide new values of some of the coefficients for the linear equations. The equations are then solved again to obtain a new trial solution and the process is repeated as necessary until convergence is achieved.

The mathematical background to the method has already been discussed in section 3.3.1. In this chapter, some applications of the method are described briefly. It should be added that the method is still under development and has not yet been fully proved as a chemical-plant-simulation technique. However, published examples and work currently in progress look very promising, particularly for use in systems with many interlocking recycles and with awkward constraints to satisfy.

In the following section the method is applied to a highly nonlinear extraction situation by way of illustration of the basic principles.

9.1.2 Extraction of halibut liver oil

Halibut liver oil (HLO) is extracted from macerated halibut livers with ether in a counter-current extraction train. For the purposes of illustration we consider a performance-type calculation involving four extraction stages, as shown in figure 9.1, in which each stream of extract (E) or liver (L) is numbered with the stage from which it comes.

In this plant, the L streams have three components – the oil-free liver solids (OFL), the retained oil and ether – while the E streams have two components only – oil and ether. For the four stages shown in figure 9.1 there are, therefore, 25 flow quantities, which may be taken as the variables for the system. The nonlinearity of the system is determined by the retention behaviour of the liver with respect to the ether and the HLO. Figure 9.2 shows, in graphical form, the retention of HLO and ether per unit mass of OFL as a function of the mass fraction of HLO in the solution (E stream) leaving a stage. While the behaviour of the ether is approximately linear, that of the HLO is certainly not.

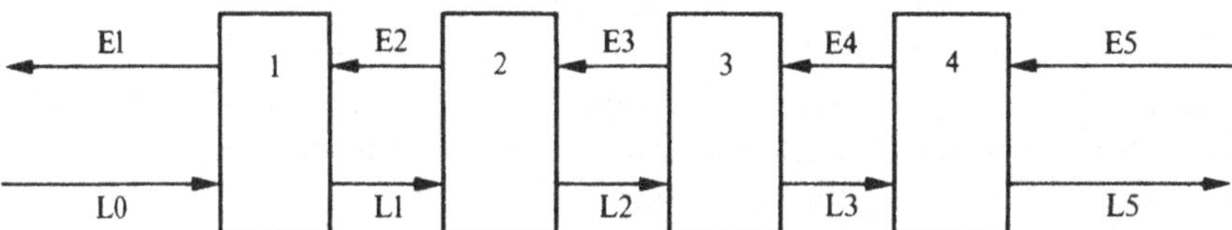

Figure 9.1. HLO (halibut liver oil) extraction train.

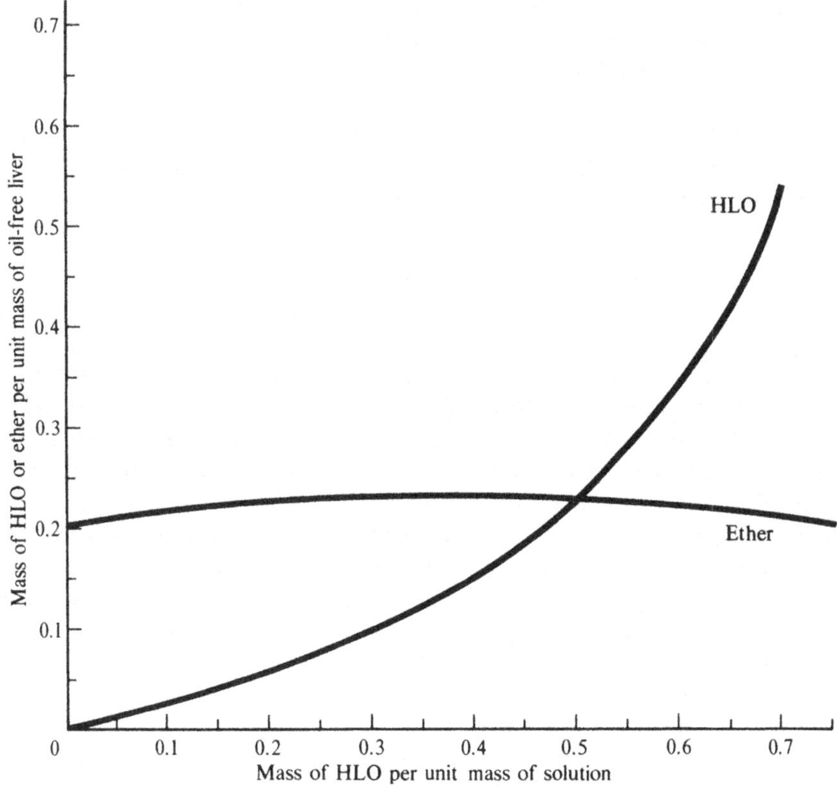

Figure 9.2. Retention factors for HLO and ether.

Each stage gives rise to five equations: there is conservation of OFL, of HLO and of ether; and there is retention of HLO and of ether by the L stream. To complete the specification of the simulation, five constraint equations are required. A typical set of such constraints might be: a scale of operation (say 1 unit of OFL), that the liver feed is free of ether, that the ether feed is free of HLO, that the fresh liver contains a known amount of HLO per unit mass of OFL, and that a desired concentration of HLO in ether is to be obtained. This set of constraints, of course, is not the only possible one, but it will serve to illustrate the solution of the problem. The equations for a particular numerical case are set out in table 9.1, in which the unknown retention factors for the HLO and ether are indicated by RH and RS respectively. The process of solution consists of guessing three compositions (that for E1 is already specified) for E2, E3 and E4, reading the eight corresponding RH and RS values from the retention graph and substituting in the appropriate equations, and solving the equations for all the flows required. Knowing these flows the compositions can be determined, and the solution process continued until the required convergence has been achieved. In fact, the retention data are not based on any very reliable experimental data, and the results shown in table 9.2 show an adequate degree of convergence to a solution.

Notice that the equations of table 9.1 could easily be reduced somewhat by algebraic manipulation. There is no advantage in doing so, however, given an adequate algorithm for the solution of sparse linear equations such as described in section 3.2.3.

9.2 Simulation of flows in pipe networks

9.2.1 Statement of the method

An example of the application of the general quasi-linear method to a situation not unrelated to simulation of a general chemical plant is provided by the simulation of flows of a single fluid, at the steady state, in a large network of pipes. In such a network, the node pressures and pipe flow velocities may be taken as the variables in a large set of equations, including conservation of mass flow equations, constitutive equations relating pressure drop to flow velocity, and constraint equations describing specified flowrates, specified node pressures and specified pump pressure drops (Bending and Hutchison, 1973).

Adopting the nomenclature indicated by figure 9.3, the following equations can be set up.

Conservation of flows at each node:

$$\sum_j \tfrac{\pi}{4} D_{ij}^2 V_{ij} + \sum_k F_k + \sum_l Q_l = 0 \tag{9.1}$$

Table 9.1. *Equations for HLO extraction train*

		L0			E1		L1			E2		L2			E3		L3			E4		L4					RHS
		OFL	HLO	Ether	HLO	Ether	OFL	HLO	Ether	HLO	Ether	OFL	HLO	Ether	HLO	Ether	OFL	HLO	Ether	HLO	Ether	OFL	HLO	Ether	HLO	Ether	RHS
Conservation	1	1					-1																				
	2		1		-1			-1		1																	
	3			1		-1			-1		1																
Retention	4				RH		-1																				
	5					RS	-1																				
Conservation	6						1					-1															
	7							1		-1			-1		1												
	8								1		-1			-1		1											
Retention	9									RH		-1															
	10										RS	-1															
Conservation	11											1					-1										
	12												1		-1			-1		1							
	13													1		-1			-1		1						
Retention	14														RH		-1										
	15															RS	-1										
Conservation	16																1					-1					
	17																	1		-1			-1		1		
	18																		1		-1			-1		1	
Retention	19																			RH		-1					
	20																				RS	-1					
Scale	21	1																									
Feed specification	22		1																								1.00
Solvent specification	23																										0.00
Feed specification	24	257	-1																						1		0.00
Extract specification	25				-1	2.333																					

Table 9.2 *Convergence behaviour of HLO extraction train calculation*

Trial no.		1		2		3		4		5	
Flow		RH or RS	Flow	RH or RS	Flow	RH or Rs	Flow	RH or RS	Flow	RH or RS	Flow
E1	HLO	0.535	0.157	0.535	0.137	0.535	0.162	0.535	0.172	0.535	0.170
	Ether	0.230	0.067	0.230	0.059	0.230	0.069	0.230	0.074	0.230	0.073
	Composition†	–	0.70	–	0.70	–	0.70	–	0.70	–	0.70
E2	HLO	0.400	0.435	0.355	0.415	0.348	0.440	0.355	0.450	0.357	0.448
	Ether	0.230	0.297	0.238	0.289	0.239	0.299	0.237	0.304	0.237	0.303
	Composition†	–	0.595	–	0.590	–	0.596	–	0.597	–	0.597
E3	HLO	0.250	0.300	0.242	0.235	0.190	0.253	0.200	0.270	0.209	0.270
	Ether	0.230	0.297	0.242	0.297	0.243	0.308	0.243	0.311	0.243	0.310
	Composition†	–	0.503	–	0.443	–	0.452	–	0.465	–	0.466
E4	HLO	0.100	0.150	0.120	0.122	0.095	0.095	0.085	0.115	0.085	0.120
	Ether	0.230	0.297	0.235	0.301	0.235	0.312	0.234	0.317	0.235	0.318
	Composition†	–	0.336	–	0.288	–	0.234	–	0.267	–	0.273
E5	HLO	–	–	–	–	–	–	–	–	–	–
	Ether	–	0.297	–	0.294	–	0.304	–	0.308	–	0.308
	Composition†	–	–	–	–	–	–	–	–	–	–
	Recovery of HLO		0.61		0.51		0.63		0.67		0.66

† Mass of HLO per unit mass of solution

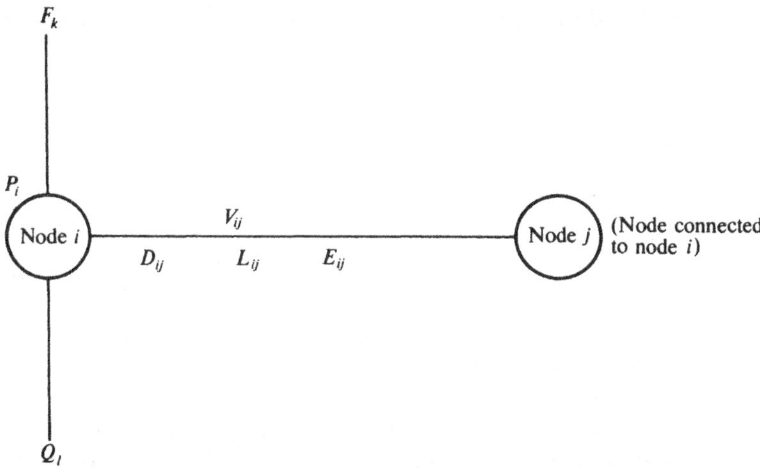

Figure 9.3. Flow between nodes i and j in a pipe network: D_{ij}, diameter of pipe; E_{ij}, roughness of pipe; L_{ij}, length of pipe; V_{ij}, superficial pipe velocity; F_k, volumetric input/output; Q_l, volumetric pump throughout; P_i, pressure at node i.

Pressure drop for each pipe:
(i) for turbulent flow

$$P_i - P_j + (4C_{ij}\rho L_{ij}/D_{ij})|V_{ij}|V_{ij} = 0 \qquad (9.2)$$

(ii) for laminar flow

$$P_i - P_j - 32\mu(L_{ij}/D_{ij}^2)V_{ij} = 0$$

Specified pressure drop for some or all pumps

$$P_i - P_l = b \qquad (9.4)$$

Specified flowrates for some or all inputs

$$F_k = b \qquad (9.5)$$

Specified flowrates for some or all pumps

$$Q_l = b \qquad (9.6)$$

Additional node pressure specifications

$$P_i = b \qquad (9.7)$$

Of these equations, only that for turbulent flow, equation (9.2), is a nonlinear constitutive equation. The conservation equation, equation (9.1), the alternative constitutive equation, equation (9.3), and the constraint equations, equations (9.4) to (9.7), are all linear. It is of course, important that the correct number of constraint equations be chosen to make the system defined.

To deal with the nonlinear equation, equation (9.2), it is sufficient to treat it as a linear equation

$$P_i - P_j + S_{ij}V_{ij} = 0 \tag{9.8}$$

with

$$S_{ij} = (4C_{ij}\rho L_{ij}/D_{ij})|V_{ij}| \tag{9.9}$$

and

$$C_{ij} = 2.5 \log_e [0.27E_{ij}/D_{ij} + 0.885/(Re\ C_{ij}^{1/2})] \tag{9.10}$$

This last equation is the Colebrook equation which takes account of the Reynolds number, Re, and the roughness of the pipe, E_{ij}, in evaluating the friction factor, C_{ij}. Note that, in itself, C_{ij} must be calculated iteratively, since it appears on both sides of equation (9.10). However since this is embedded in an iterative process anyway, it proves to be sufficient simply to do one stage of improvement of C_{ij} with each iteration of the main solution process.

The solution process, then, is to set up the equations (9.1), (9.8), and (9.4)–(9.7) for an assumed, usually uniform, pipe velocity in the turbulent region. The equations are then solved for the node pressures and the pipe velocities, the coefficients for equations (9.8) recalculated, or if appropriate substituted with equation (9.2), and the solution process repeated until a converged solution is achieved. Unfortunately, this simple scheme does not converge very rapidly: the node pressures settle down rapidly but the pipe velocities oscillate about their correct values. The reason for this is easily seen from elementary considerations. Suppose the node pressures have settled, approximately, to their final values. Then the turbulent equations have the form

$$\Delta P = K|V^{(p)}|V^{(p+1)} \tag{9.11}$$

where the superscript p is an index of iteration. Ignoring the small dependence of K on pipe velocity

$$V^{(p+1)} = \Delta P/K|V^{(p)}| \tag{9.12}$$

and

$$V^{(p+2)} = \Delta P/K|V^{(p+1)}| = V^{(p)} \tag{9.13}$$

Two stages of iteration therefore appear to bring the calculation back to where it started from. This situation is relieved by relaxing the newly computed pipe velocity with the old one, i.e.

$$V_{ij}^{(p+1)} = 0.5(V_{ij}^* + V_{ij}^{(p)}) \tag{9.14}$$

where V_{ij}^* is the pipe velocity obtained by solution of the set of linear equations.

This procedure turns out to be very satisfactory: a number of networks comprising up to 38 pipes with 22 nodes have been calculated, most converging in under eight iterations, except for mixed laminar–turbulent situations which require 10–12 iterations for convergence. There appears to be no inherent limitation for size, given efficient techniques for solving sparse equations.

9.2.2 Further consideration of the convergence

In this section, which may be omitted on first reading, further consideration is given, in the light of section 3.3.1, to the convergence of this iterative procedure. Attention is focused on the representation of the nonlinear equations (9.2), which we here simplify (disregarding the dependence of C_{ij} on the flowrate) to

$$P_i - P_j + K|V_{ij}|V_{ij} = 0 \tag{9.15}$$

The scheme of iteration which gives rise to the oscillatory, and not very rapid, convergence is to represent this equation by

$$P_i - P_j + \{K|V_{ij}|\}^{(p)}V_{ij}^{(p+1)} = 0 \tag{9.16}$$

in which the quantity within the braces is evaluated using an existing estimate of the flowrate. This is entirely equivalent to the scheme derived from equations (3.20). An alternative scheme would be to represent equation (9.15) by

$$P_i - P_j + 2\{K|V_{ij}|\}^{(p)}V_{ij}^{(p+1)} = \{K|V_{ij}|\}^{(p)}V^{(p)} \tag{9.17}$$

which is the application to equation (9.15) of the general result of equation (3.30). Sufficiently close to the solution, the relationship between the two schemes may be seen to be as shown in figure 9.4. In this figure a Newton–Raphson-type extrapolation to B gives an answer very close to the true solution B', while extrapolation to C (which is the equivalent of the process of equation (9.15)) with a slope one half of the extrapolation to B, yields a less satisfactory answer. However, relaxing C by taking a new solution to be one half the old solution A, and one half the new C, also leads to the same value B. Close enough to the solution the method of equation (9.17) and the method of relaxing the results of equation (9.16) appear to be very nearly identical.

In this particular case, which is complicated by the non-linear dependence of K on V_{ij} in equation (9.15), it is sufficient to adopt the method of equation (9.16) with relaxation, rather than go to the more cumbersome equation (9.17).

9.3 Application to distillation

The original development of the linear equation approach to distillation

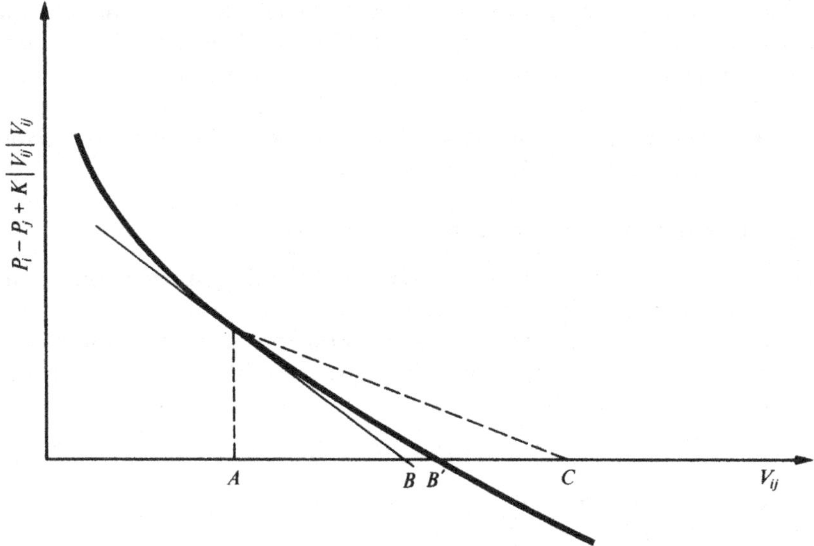

Figure 9.4. Comparison of Newton step with quasi-linear step.

was by Amundson (1966). However, the methods that were advocated by him and by subsequent developers depended on partitioning the whole system of equations by component, a partitioning that was induced, if not rendered essential, by the limitations of size of computers then available. By contrast, the method outlined here treats all the components simultaneously. Further, it is possible to handle the calculations for several columns simultaneously, as will be discussed below. The method here described is essentially that of Hutchison and Shewchuk (1974).

9.3.1 General description of the method

In the method to be described, it is presumed that the hardware configuration of a separation column or group of columns has been defined, together with the components which represent the substances presumed present in the system. The explicit variables of the system are taken to be all the flowrates (mole units usually) and all the compositions of the components in every stream in the system; streams will include both obvious physical streams and also, for example, the counter-current streams passing between stages internal to a column. This complete set of variables for the whole system is treated as the set of variables for a single set of equations. There will be conservation equations, such as the linear equations for conservation of mass, and the nonlinear equations for conservation of energy. There will be constitutive equations, such as the vapour–liquid equilibrium relationships discussed in the next section, and there will be constraint equations defining the operating conditions for the column or columns.

Many of the coefficients of the variables are constants, but some are dependent upon temperature, pressure, and the composition of the stream to which they relate, and these coefficients must be calculated using the best available estimate of the appropriate stream quantities. Solution of the linear equations so set up leads to a new estimate of the stream quantities, from which revised estimates of the coefficient values may be obtained. In most situations the iterative process so described converges quite rapidly, even for substantial systems.

The conservation equations for a single stage require no further explanation. They are the familiar component and enthalpy balances for single stages which have often been discussed. They may be extended to account for stage efficiency and for entrainment without difficulty. The constitutive equations for vapour–liquid equilibrium, however, may take an unusual form, as discussed in the next section, with consequential improvement in the rate of convergence of the calculation.

9.3.2 Vapour–liquid equilibrium

The conventional treatment of vapour–liquid equilibrium in the context of multicomponent distillation is to use an equation

$$y_i = K_i x_i \qquad (9.18)$$

where K_i is a function of temperature and pressure, and (may be) also of composition. It is certainly possible to use such a formulation in the distillation method being described, but an improved formulation which leads to a substantial improvement in the rate of convergence is also possible. Because all the compositions of both liquid and vapour on a stage are treated as variables in the same set of equations, it is possible to use an equation

$$y_i = \sum_j K_{ij} x_j \qquad (9.19)$$

where the summation is carried over all the liquid compositions on the stage. Qualitative justification for such a representation may be obtained by considering a binary system, as in figure 9.5.

It is immediately clear that the line $D_1 D_2$ is a better representation of the equilibrium curve relating y_B to x_B, in the neighbourhood of P, than is the line AC. The distance BC represents the K_B of equation (9.18), while the distances BD_1 and AD_2 represent the K_{BB} and K_{BA} of equation (9.19). For constant relative volatility a in a multicomponent system it may be shown that

$$K_{ij} = \frac{a_i(a_{av} - a_j)x_i}{a_{av}^2} + \frac{a_i \delta_{ij}}{a_{av}} \qquad (9.20)$$

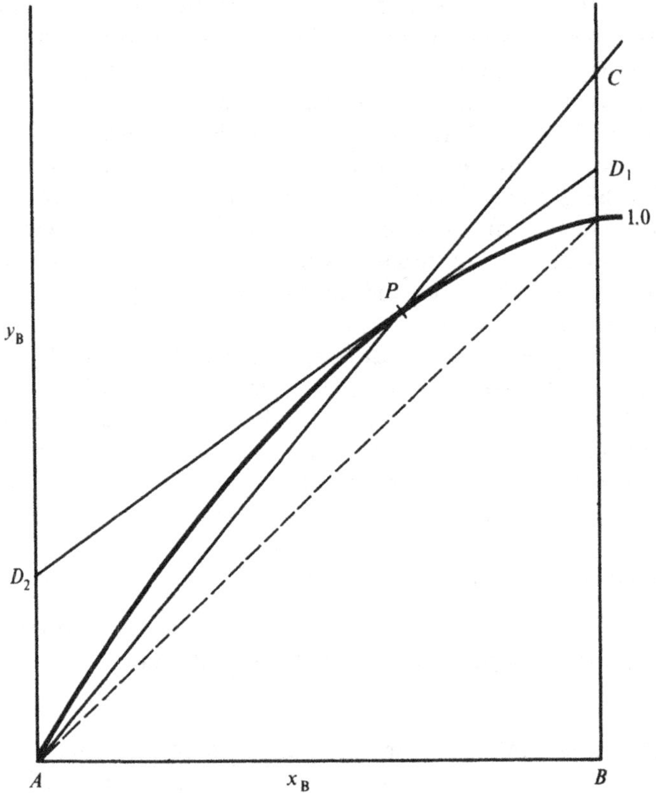

Figure 9.5. Equilibrium curve for binary system. Lines AC and D_2D_1 represent two different linearization schemes through point P.

where

$$a_{av} = \sum_i a_i x_i \qquad (9.21)$$

and δ_{ij} is the Kronecker delta of i and j. For less ideal systems more complicated equations can be derived relating K_{ij} to the relative volatilities and to the activity coefficients, and thus indirectly to the compositions, in a manner analogous to equations (9.20).

The more mathematically minded reader may wish to consider this treatment of vapour–liquid equilibrium in the light of equations (3.32) to (3.33).

The hyper-plane treatment of vapour–liquid equilibrium contained in equation (9.19) has been shown to give rise to second-order convergence in several systems for which equation (9.18) gave rise only to first-order convergence, with consequential increase in computing time.

9.3.3 Segmentation of distillation columns

For the purposes of setting up the quasi-linear equations automatically, it is convenient to segment a distillation column into a number of sections. These sections are standard sections, for example, a column of stages, or a partial condenser, which may be put together in any desired fashion to represent a single column.

Table 9.3 gives a list of typical column elements.

Table 9.3 *Typical column elements*

Column of stages	A column of m stages, which may be theoretical stages, or with an efficiency to be defined, and/or entrainment to be defined. Feed, sidestream withdrawal, and heat exchange may be provided on each stage. m must be predetermined
Equilibrium boiler	Representing a kettle boiler
Total boiler	Representing a total boiler
Total condenser	Representing a total condenser
Partial condenser	Representing a single equilibrium stage of a partial condenser
Flow divider	Representing a reflux or reboil divider, or any stream division
Mixer	Representing mixing of streams
Temperature setters	Representing heaters or coolers controlled to defined outlet temperatures
Heater/cooler	Representing specific heating/cooling loads
Heat exchangers	Representing heat interchange between streams
Stream junctions	These represent the junction of two streams, and serve to indicate the identity of a stream leaving a column section with that entering a condenser section etc.
Stream constraints	These represent the operating conditions imposed upon the column

9.3.4 Simulation of separation trains

Given the availability of standard column segments, as described in the last section, it is possible to simulate individual towers of arbitrary complexity, or complete separation plants, without further difficulty. For example, the column shown in figure 9.6, with reflux provided by cold liquid drawn from stage 13 and returned to stage 15, may be represented by units (drawn from table 9.3) consisting of: a column of stages, with one feed and one liquid sidestream; a temperature setter to represent the sidestream cooler; an equilibrium boiler; and a number of junctions (indicated by solid circles on the figure). Such a column has been simulated using a six-component mixture of light hydrocarbons operating at 3 bar, and required some 16 iterations to converge to a satisfactory solution. Another interesting simulation achieved by this method is of the Petlyuk towers (Petlyuk *et al.*, 1965) shown in figure 9.7, in which a three-component mixture is separated into almost pure components using two columns with a single total reboiler and total condenser. This case may be discussed in more detail by indicating the information required to define the simulation.

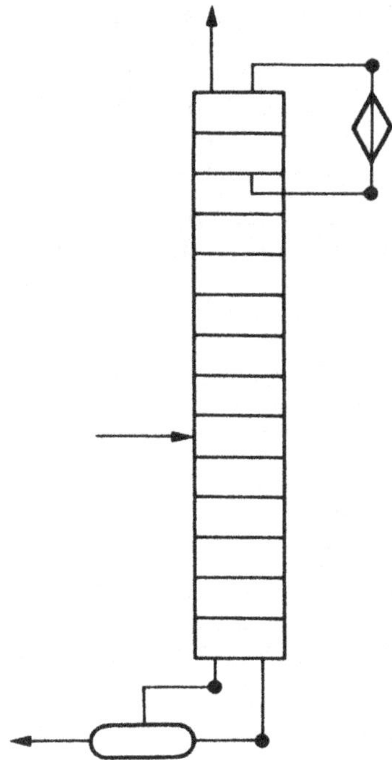

Figure 9.6. Debutanizer with pump-around circuit.

(*a*) Tower 1 has fifteen stages with feeds to stages 4 and 12. Liquid sidestreams are taken from stages 8 and 13. A vapour sidestream is taken from stage 3.

(*b*) Tower 1 is supplied with a total condenser, followed by a flow divider which prescribes the reflux to distillate ratio.

(*c*) Tower 1 has a total reboiler, preceded by a flow divider which allows a bottom product to be removed at a specified rate.

(*d*) Tower 2 has seven stages with a feed to stage 4.

(*e*) Junctions are provided between the liquid from stage 1 of tower 2 and the feed to stage 4 of tower 1, between the sidestream from stage 3 of tower 1 and the vapour to stage 1 of tower 2, and similarly at the top of tower 2, and also around the condenser and reboiler.

(*f*) The flowrates of the sidestreams from stages 3, 8 and 13 are specified.

(*g*) The composition, enthalpy and flowrate of the feed to tower 2 is completely specified.

(*h*) The operating pressure (assumed uniform) is specified.

The mixture used for the simulation of these towers was n-hexane, n-octane and n-decane, and ideal behaviour was assumed, relative volatilities at the column temperatures being generated from Antoine

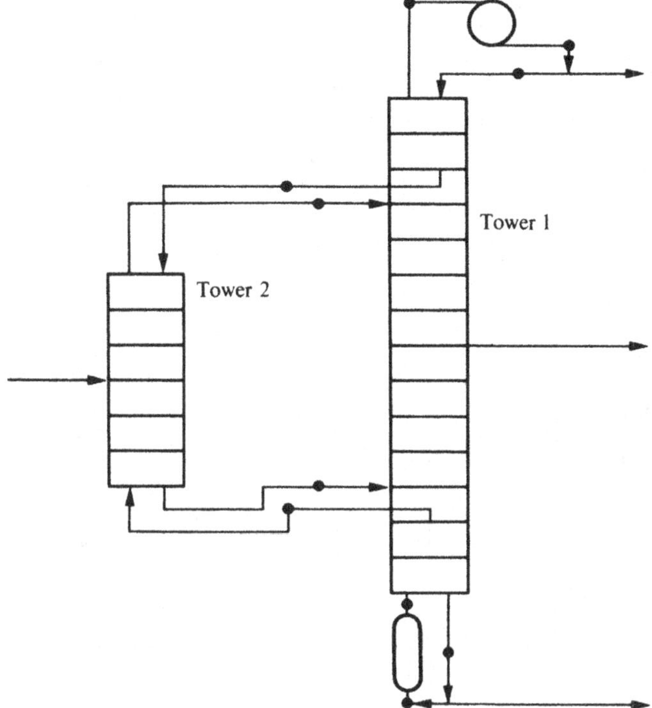

Figure 9.7. Petlyuk towers.

equations. All stages were assumed to be ideal. A converged solution
was obtained on the ninth iteration.

For interest, this system requires 300 equations in 300 variables, the
equations having 1070 nonzero terms. The solution process, by the
methods of section 3.2, generated 503 additional nonzero terms. The
total number of nonzero terms stored for these equations, using the
methods of section 3.2, were thus less than 2% of the total number of
terms, 90 300, which would have been required for full storage of the
matrix of equations.

A further example of a system of columns simulated by this technique
is shown in figure 9.8. It involves three separate distillation or absorption columns with two major recycle loops; treating a five-component
mixture, convergence is achieved in 15 iterations of the whole plant
calculation.

At this stage one may well ask how much further it is necessary to go
to achieve a process plant simulation. This theme will be taken up in
section 9.5. First follows a brief description of the application of the
method to multiple reaction equilibrium.

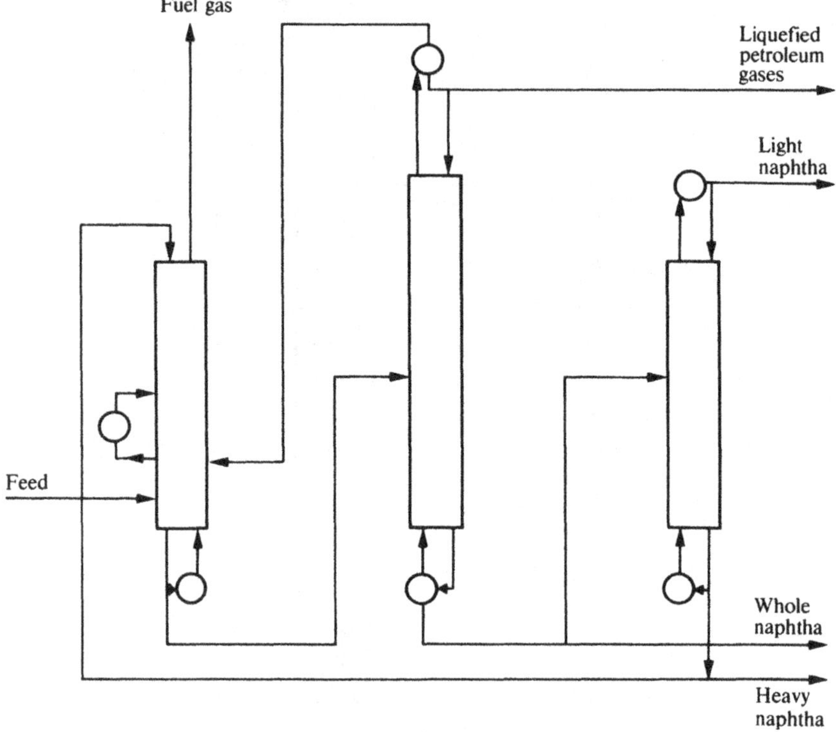

Figure 9.8. Simplified naphtha separation plant.

9.4 Application to multiple reaction equilibrium

This section is included as an illustration in its own right of the use of the quasi-linear method, but also, as will be discussed in section 9.4.4, as an illustration of the methods of blending specific reactor models into the general concept of quasi-linear simulation.

9.4.1 Multiple reaction equilibrium; chemical potentials

We consider a situation in which a stream of material is brought to chemical equilibrium, there being several independent reactions among the components of the mixture, all of which reactions are presumed to be at equilibrium, according to the principle of detailed balancing.

For each reaction, formally described in terms of its stoichiometric coefficients (equation (5.15)),

$$\sum_i v_{ir} M_i = 0 \tag{9.22}$$

the reaction equilibrium condition is (equation (5.19))

$$\sum_i v_{ir}\,\mu_i = 0 \tag{9.23}$$

which is a linear equation in the chemical potentials, μ_i. Since each chemical species can be expressed in terms of its atomic groups (atoms or groups of atoms which are unchanged in the course of any of the reactions taking place)

$$M_i = \sum_i \beta_{ij} A_k \tag{9.24}$$

and since the atomic groups A_i are by definition conserved, we have k conservation conditions,

$$\sum_i \beta_{ik}\eta_i = \eta_k^0 \tag{9.25}$$

where η_i is the number of moles of species i present at any time, and η_k^0 is the total number of moles (so to speak) of the atomic groups present in the system. It is an elementary result of the phase rule that the number of reaction equilibrium conditions, equations (9.23), and the number of atomic group conservation equations, equations (9.25), are together equal to the number of components present. We therefore take the number of moles of each component present, and the chemical potential of each component, as variables in a set of $2c$ equations where c is the total number of components. Of these equations c are covered by c linear equations of the form (9.23) or (9.25), and the other c are the nonlinear constitutive equations relating the chemical potentials to the composition of the reacting mixture:

$$\mu_i - f_i(\eta_i, \ldots, \eta_c) = 0 \tag{9.26}$$

Because of the Gibbs–Duhem relationship, this equation can be linearized in the form (see equation (3.30))

$$\mu_i - \sum \eta_j f_{ij}(\eta_2, \ldots, \eta_c) = -f_i(\eta_1, \ldots, \eta_c) \tag{9.27}$$

where f_{ij} is the partial derivative of f_i with respect to η_j. In terms of an iterative process, an estimated set of mole numbers $\eta_i^{(p)}$ is used to evaluate the $f_{ij}(\eta_i^{(p)}, \ldots, \eta_c^{(p)})$ and $f_i(\eta_i^{(p)}, \ldots, \eta_c^{(p)})$, so setting up a set of linear relationships between μ_i and the η_i (c equations). These are combined with equations (9.23) and (9.25) to form a set of $2c$ equations, from which a fresh estimate of the mole numbers can be established. For example, for an ideal gas mixture

$$\left. \begin{aligned} f_i &= \mu_i^0 + RT \ln p_i \\ &= \mu_i^0 + RT \ln P + RT \ln \eta_i - RT \ln \sum \eta_i \\[2mm] f_{ij} &= \frac{RT\delta_{ij}}{\eta_i} - \frac{RT}{\sum \eta_i} \end{aligned} \right\} \tag{9.28}$$

and

9.4.2 Application of the method to a specific case

Consider the application to the combustion of methane in a limited supply of oxygen, so that the possible components in the reaction mixture are CH_4, H_2, H_2O, CO, CO_2. The amount of free oxygen is so small as to be neglected. Solid carbon is not formed unless the oxygen is reduced too far.

Between these five species there are two independent reactions, which may be taken to be

$$\left. \begin{array}{l} H_2 + CO_2 \rightleftharpoons CO + H_2O \\ CH_4 + H_2O \rightleftharpoons CO + 3H_2 \end{array} \right\} \qquad (9.29)$$

These give rise to the equilibrium conditions

$$\left. \begin{array}{l} \mu_{H_2} + \mu_{CO_2} - \mu_{CO} - \mu_{H_2O} = 0 \\ \mu_{CH_4} + \mu_{H_2O} - \mu_{CO} - 3\mu_{H_2} = 0 \end{array} \right\} \qquad (9.30)$$

where for convenience, and in subsequent usage in this section, each chemical potential is divided by RT.

The conservation equations are

$$\left. \begin{array}{l} \eta_{CH_4} + \eta_{CO} + \eta_{CO_2} = \eta_C^0 \\[2ex] 4\eta_{CH_4} + 2\eta_{H_2} + 2\eta_{H_2O} = \eta_H^0 \\[2ex] \eta_{H_2O} + \eta_{CO} + 2\eta_{CO_2} = \eta_O^0 \end{array} \right\} \qquad (9.31)$$

for carbon,

for hydrogen, and

for oxygen.

The remaining five equations have the form

$$\mu_{CH_4} - \left(\frac{1}{\eta_{CH_4}} - \frac{1}{\Sigma\eta} \right)^{(p)} \eta_{CH_4} + \left(\frac{1}{\Sigma\eta} \right)^{(p)} (\eta_{H_2} + \eta_{H_2O} + \eta_{CO} + \eta_{CO_2})$$
$$= \mu_{CH_4} + \ln P + \ln \eta_{CH_4}^{(p)} - \ln(\Sigma\eta)^{(p)} \qquad (9.32)$$

We also have

$$\Sigma\eta = \eta_{CH_4} + \eta_{H_2} + \eta_{H_2O} + \eta_{CO} + \eta_{CO_2} \qquad (9.33)$$

Consider the particular case of the combustion of 1 mole of CH_4 with 0.505 moles O_2, at 1 bar, and at a temperature such that the modified standard chemical potentials are as in table 9.4. If the trial values shown in the first column of table 9.5 are adopted, the equations shown in table 9.6 arise, from which it can easily be verified that the results shown in the second column of table 9.5 may be obtained. The last column of table 9.5 shows the resulting chemical potential of the components in the equilibrium mixture, as they are obtained from the same set of equations.

Table 9.4 *Standard chemical potentials,*
μ^0/RT

CH_4	$+0.85$
H_2	$0.$
H_2O	-23.40
CO	-23.50
CO_2	-47.00

Table 9.5. *Trial values and first solution*

	Trial	Solution	μ/RT
CH_4	0.308	0.3092	-1.1914
H_2	1.180	1.1748	-0.7066
H_2O	0.205	0.2068	-25.8071
CO	0.580	0.5783	-24.8881
CO	0.113	0.1125	-50.1854
$\Sigma\eta^2$	2.386	2.3815	

9.4.3 Multiple reaction equilibrium; alternative method

An alternative method for tackling the same problem makes use of equilibrium constants in place of the chemical potentials to describe the thermodynamic information. In this method, which is illustrated with the same example as was used in the previous section, equations (9.31) (three equations representing conservation of mass) and (9.33) (summation of mole numbers) are taken over directly. Two more equations, equations (9.34), are provided by the equilibrium relationships corresponding to the reactions (9.29).

$$\left.\begin{aligned} K_1\eta_{H_2}\eta_{CO_2} - \eta_{CO}\eta_{H_2O} = 0 \\ K_2\eta_{CH_4}\eta_{H_2O}(\Sigma\eta)^2 - \eta_{CO}(\eta_{H_2})^3 = 0 \end{aligned}\right\} \tag{9.34}$$

There are therefore six equations in the five mole numbers η and their sum $\Sigma\eta$, of which two, as above, are nonlinear.

Following the method set out in section 3.3.3 these equations may be linearized as in equation (9.35) in which terms precalculated using existing values are indicated in parentheses.

$$\left.\begin{aligned} (K_1\eta_{H_2})\eta_{CO_2} &+ (K_1\eta_{CO_2})\eta_{H_2} - (\eta_{CO})\eta_{H_2O} - (\eta_{H_2O})\eta_{CO} \\ &= (K_1\eta_{H_2}\eta_{CO_2} - \eta_{CO}\eta_{H_2O}) \\ (K_2\eta_{CH_4}(\Sigma\eta)^2)\eta_{H_2O} &+ (K_2\eta_{H_2O}(\Sigma\eta)^2)\eta_{CH_4} + (2K_2\eta_{CH_4}\eta_{H_2O}\Sigma\eta)\Sigma\eta \\ &- (\eta_{H_2})^3\eta_{CO} - (3\eta_{H_2}^2\eta_{CO})\eta_{H_2} = 3(K_2\eta_{CH_4}\eta_{H_2O}(\Sigma\eta)^2 - \eta_{CO}\eta_{H2}^3) \end{aligned}\right\} \tag{9.35}$$

Table 9.6 *Equations for solution of reaction equilibrium*

μ_{CH_4}/RT	μ_{H_2}/RT	μ_{H_2O}/RT	μ_{CO}/RT	μ_{CO_2}/RT	η_{CH_4}	η_{H_2}	η_{H_2O}	η_{CO}	η_{CO_2}	$\Sigma\eta$	RHS
	1	−1	−1	1							
1	−3	1	−1								
					1			1	1		1.00
					4	2	2				4.00
							1	1	2		1.01
1					−2.82764	0.41911	0.41911	0.41911	0.41911		−1.19727
	1				0.41911	−0.42835	0.41911	0.41911	0.41911		−0.70410
		1			0.41911	0.41911	−4.45894	0.41911	0.41911		−25.85430
			1		0.41911	0.41911	0.41911	−1.30503	0.41911		−24.91435
				1	0.41911	0.41911	0.41911	0.41911	−8.43055		50.04999
					1	1	1	1	1	−1	

Using the same conditions and starting values as in the previous example, and evaluating K_1 and K_2 in the usual way to be 0.90484 and 2.58571 respectively, the equations shown in table 9.7 may be obtained, the solution to which does not differ substantially from that obtained in table 9.5, although it does not give the chemical potentials at the same time.

Table 9.7 *Alternative linear equations for reaction equilibrium*

η_{CH_4}	η_{H_2}	η_{H_2O}	η_{CO}	η_{CO_2}	$\Sigma\eta$	RHS
1	1	1	1	1	-1	
1			1	1		1.000
4	2	2				4.000
		1	1	2		1.010
	0.10225	-0.580	-0.205	1.06771		0.00175
3.01769	-2.42278	4.53389	-1.54303		4.48449	-0.10664

9.4.4 Reaction equilibrium in relation to process simulation

It is clear from the preceding sections that the overall philosophy in dealing with equilibrium in multiple reaction systems is very similar to that adopted in the multiple column program: a set of equations is developed comprising conservation equations, constitutive equations and constraints, the nonlinear equations are linearized in some fashion, and the whole set solved iteratively until convergence is achieved. There is therefore little difficulty in blending an equilibrium reactor of this character into a general quasi-linear flowsheeting system of the same basic characteristics.

There is, of course, more than one method of achieving the linearization. The method selected will have an important effect on two questions: how rapidly will the iteration process converge once it gets properly started, and how near the solution does one have to guess the starting values in order for the iteration process to converge at all? We may characterize the answers to these two questions as 'efficiency' and 'robustness' of convergence respectively. Efficiency can be achieved by attaining second-order convergence, as by a Newton–Raphson scheme, but only at the expense of robustness. Robustness can be achieved using first-order methods, but these may be slow away from the solution. Since second-order methods and first-order methods differ only in the numerical values which are used in the quasi-linear equations, it is possible easily to switch from one to the other, or to blend the two together, as the convergence proceeds. There have been some experiments along these lines, and, for a single unit calculation such as multiple reaction equilibrium, progressive blending of methods, by changing the value of a single parameter of a linear combination, appears to be successful. Further work, however, is required along these lines.

9.5 Towards process simulation by quasi-linear methods

The culmination of the ideas that have been developed in this chapter would be a full process simulation by quasi-linear methods. This appears not to have been attempted at present, although research along these lines is continuing, and success appears at present only a matter of time and development effort. This section indicates briefly some of the characteristics of such a general-purpose quasi-linear process simulator, and may in some sense act as a summary to this chapter.

9.5.1 General characteristics

The general features of such a system are now plain. It will be based on an efficient method for solving sparse matrix equations, with a built-in facility for achieving second and subsequent solutions of the same set of equations, with only the values of the equations changed, much more rapidly than the primary solution can be achieved.

The unit operations of the chemical engineer are preserved, but the essential character of the subroutines which embody them is substantially altered. Such routines are likely to be in three main parts, apart from error-handling sections: the first part sets up the appropriate linear equations for the unit operation, the second part is concerned with evaluating the latest value of the terms in the linear equations, and the third part is concerned with any subsequent calculations and reports which may be necessary once a converged solution has been achieved.

The capability of enforcing constraints on a solution, which is such an important feature of the strictly linear process simulation technique, is fully preserved in the quasi-linear simulator. Demand-dominated calculations, explicit composition requirements, and similar constraints pose no difficulties to the quasi-linear simulator.

It is clear, also, that the quasi-linear simulator can deal just as easily with a highly convoluted system with many information recycles as with simple systems with a single recycle. It is not at its most efficient when dealing with a strict feed-forward system of unit calculations. Such a system is best handled as a serial computation, even if quasi-linear methods are used for some of the unit calculations. But as systems become more and more 'cross-linked' the advantage of the quasi-linear method becomes more apparent.

9.5.2 Areas in which further development is necessary

There are some areas in which further development or research is necessary. The most fundamental concerns the choice of the method of linearization for individual unit operations to achieve a satisfactory blend between robustness and efficiency of convergence. It would be of great practical value to have realistic estimates of the dimensions of the hyper-ellipsoid within which second-order methods would converge to a particular solution, even if slowly to begin with. It would also be useful to have realistic estimates of the dimensions of the hyper-ellipsoid within which a given first-order method would converge. Work must also be done on the methods of blending first- and second-order methods. Should this be achieved by means of a single parameter for the whole process simulation, or would it be better to allow for separate blending parameters for each of the unit operations of the chemical engineer?

Considerable work must go into developing further unit operations for use with a quasi-linear simulator. This is an area which hopefully will be taken up by the industrial users of such a system once they are convinced of the general usefulness of the system. It may be that further work on the thermodynamic methods used will reveal that there are advantages in computational time and cost to be gained by 'breaking open' the thermodynamic packages and submitting these to a process of linearization also.

Finally, although the techniques now available for handling sparse matrix equations are capable of handling one or two thousand variables (perhaps even more), realistic simulation of chemical plant is going to require, within the next decade, very many more variables. Beyond chemical engineering, too, in economics and ecology, the need for simulation of large systems is apparent. Work must be done on developing sparse matrix methods to work on equation sets larger by an order of magnitude or more. With methods at his fingertips for simulating systems of 20 000 variables, in both steady and dynamic states, the chemical engineer of tomorrow will be better able than his counterpart of today to deal with the problems which an engineer is called upon to face.

10

Further reading and literature references

Several review articles will lead one to most of the literature occurring before 1976 in the area of computer-aided design. A most extensive review by Hlavacek (1977) lists over 400 references in the areas of reconciliation of plant data, steady-state and transient simulation and process synthesis. Motard *et al.* (1975) have over 100 flowsheeting-oriented references covering the period 1970 to 1974. Two review articles, Flower and Whitehead (1973*a, b*), and Evans *et al.* (1968), list many of the flowsheeting programs mentioned throughout the literature. Mah (1974) has reviewed much of the literature on structural analysis of equations and flowsheets as well as various approaches used for computation in flowsheeting systems. Kehat and Shacham (1973*a, b, c*) review much of the flowsheeting literature up to 1973, giving a set of general classifications as to the types of systems. The earliest review known was by Sargent (1967) in which he discusses all aspects of computer-aided design and presents a list of the literature up to that time.

A review article in the related area of process synthesis was written by Hendry *et al.* (1973). The literature has grown rapidly in this area since that time, with most of the articles appearing in the traditional chemical engineering journals.

A most extensive review with about 600 references in the sparse matrix literature is that of Duff (1976) with additional comments in a related review by Reid (1976). A number of computer programs using sparse-matrix methods are available in the Harwell subroutine library (1973); this library now contains many newer routines than listed in their 1973 list.

Amundson, N. R. (1966) *Mathematical Methods in Chemical Engineering – Matrices and their Application*, pp. 170–6. Prentice-Hall, Englewood Cliffs, New Jersey.

Barkley, R. W. and R. L. Motard (1972) Decomposition of nets. *Chem. Eng. J.* **3**, 265.

Bending, M. J. and H. P. Hutchison (1973) The calculation of steady state incompressible flow in large networks of pipes. *Chem. Eng. Sci.* **28**, 1857–64.

Broyden, C. G. (1965) A class of methods for solving nonlinear simultaneous equations. *Math. Comput.* **19**, 577.

Broyden, C. G. (1970) The convergence of a class of double-rank minimization algorithms. *J. Inst. Math. Appl.* **6**, 76–90.

CADC (1973*a*) *CONCEPT Mark III User Manual*. Computer Aided Design Centre, Cambridge, England.

CADC (1973*b*) *SYMBOL User Manual*. Computer Aided Design Centre, Cambridge, England.

CADC (1977) *MULTICOL User Manual*. Computer Aided Design Centre, Cambridge, England.

Carnahan, B., H. A. Luther and J. O. Wilkes (1969) *Applied Numerical Methods*, p. 319. John Wiley, New York.

Cavett, R. H. (1963) Application of numerical methods to the convergence of simulated processes involving recycle loops. *Am. Pet. Inst. Repr. No. 04–63*.

Cheung, L. K. and E. S. Kuh (1974) The bordered triangular matrix and minimal essential sets of a diagraph. *IEEE Trans. Circuits Syst.* **CAS-21**, 633–9.

Christensen, J. H. (1970) The structuring of process optimization. *AIChE J.* **16**, 177.

Christensen, J. H. and D. F. Rudd (1969) Structuring design computations. *AIChE J.* **15**, 94–100.

Crowe, C. M., A. E. Hamielec, T. W. Hoffman, A. I. Johnson, D. R. Woods and P. T. Shannon (1971) *Chemical Plant Simulation*. Prentice-Hall, Englewood Cliffs, New Jersey.

Crowe, C. M. and M. Nishio (1975) Convergence promotion in the simulation of chemical processes – the general dominant eigenvalue method. *AIChE J.* **21**, 528–33.

Davy Computing (1969) *COMPAID*. Davy Computing Ltd., Sheffield, England.

Dean, J. A. (ed.) (1973) *Lange's Handbook of Chemistry*, 11th edn. McGraw-Hill, New York.

deBrosse, C. J. and A. W. Westerberg (1973). A feasible point algorithm for structured design systems in chemical engineering. *AIChE J.* **19**, 251.

Dennis, J. R. Jr and J. J. More (1977) Quasi-Newton methods, motivation and theory. *SIAM Rev.* **19**, 46–89.

Digital Systems (1971) *PACER 245 User Manual*. Digital Systems Corp., Hanover, New Hampshire.

Duff, I. (1977) A survey of sparse matrix research. *Proc. IEEE*, **65**, 500–35.

Duff, I. and J. K. Reid (1976) An implementation of Tarjan's algorithm for the block triangularization of a matrix. *AERE Harwell Report CSS-29*. Atomic Energy Research Establishment, Oxfordshire, England.

Edie, F. C. and A. W. Westerberg (1971) Computer aided design. III. Decision variable selection to avoid hidden singularities in resulting recycle calculations. *Chem. Eng. J.* **2**, 114.

Evans, L. B., D. G. Steward and C. R. Sprague (1968) Computer-aided chemical process design. *Chem. Eng. Prog.* **64**, 39–46.

Flower, J. R. and B. D. Whitehead (1973*a*) Computer-aided design. A survey of flowsheeting programs: I. *Chem. Eng.* (London) **272**, 208.

Flower, J. R. and B. D. Whitehead (1973*b*) Computer-aided design. A survey of flowsheeting programs: II. *Chem. Eng.* (London) **273**, 271.

Forder, G. J. and H. P. Hutchison (1969) The analysis of chemical plant flowsheets. *Chem. Eng. Sci.* **24**, 771–85.

Fox, L. (1964) *An Introduction to Numerical Linear Algebra*. Clarendon Press, Oxford, England.

Guardabassi, G. (1974) An indirect method for minimal essential sets. *IEEE Trans. Circuits Syst.* **CAS-21**, 14.

Gupta, P. K. (1972) Application of analysis and optimization techniques for structured systems to design of a double effect evaporator system, MS thesis, University of Florida, Gainesville, Florida.

Gupta, P. K., A. W. Westerberg, J. E. Hendry and R. R. Hughes (1974). Assigning output variables to equations using linear programming. *AIChE J.* **20**, 397–9.

Guthrie, K. M. (1969) Capital cost estimating. *Chem. Eng.* (New York), **76**, 114–42, March 24.

Harary, F. (1959) A graph theoretic method for the complete reduction of a matrix with a view toward finding its eigenvalues. *J. Math. Phys.* **38**, 104–11.

Harwell subroutine library (1973) *AERE Harwell Report R-7477*. Atomic Energy Research Establishment, Oxfordshire, England.

Hendry, J. E., D. F. Rudd and J. D. Seader (1973) Synthesis in the design of chemical processes. *AIChE J.* **19**, 1–15.

Himmelblau, D. M. (1973) Morphology of decomposition. In *Decomposition of Large-Scale Problems* (ed. D. M. Himmelblau), pp. 1–13. North Holland, Amsterdam.

Hlavacek, V. (1977) Analysis and synthesis of complex plants – steady state and transient behaviour. *Comput. Chem. Eng.* **1**, 25.

Hoffman, T. W. and H. P. Hutchison (1975) The simulation of the Bayer process for extracting alumina from bauxite ore. In *Symposium on Computers in the Design and Erection of Chemical Plants*, pp. 451–85. Karlovy Vary, Czechoslovakia.

Hutchison, H. P. (1974) Plant simulation by linear methods. *Trans. Inst. Chem. Eng.* **52**, 287–90.

Hutchison, H. P. and C. F. Shewchuk (1974) Computational method for multiple distillation towers. *Trans. Inst. Chem. Eng.* **52**, 325.

ICI (1967) *ISOPEDAC*. Imperial Chemical Industries Ltd., Central Management Services, Wilmslow, England.

IHI (1975) *IPAC*. Ishikawajimi-Harima Heavy Industries Co., Ltd., Japan.

Janicke, W. and G. Biess (1974) *Chem. Technol.* **26**, 740.

Johns, W. R. (1970) Mathematical considerations in preparing general-purpose computer programs for the design or simulation of chemical processes. *EFCE Conference on the Use of Computers in Studies Preceding the Design of Chemical Plants*, Florence, Italy, April.

Johnson, A. I. (1974) *SACDA*. University of Western Ontario, London, Ontario, Canada.

Kehat, E. and M. Shacham (1973*a*) Chemical process simulation programs: I. *Process Technol.* **18**, 35.

Kehat, E. and M. Shacham (1973*b*) Chemical process simulation programs. II. Partitioning and tearing system flowsheets. *Process Technol.* **18**, 115.

Kehat, E. and M. Shacham (1973*c*) Chemical process simulation programs. III. Solution system of nonlinear equations. *Process Technol.* **18**, 181.

Kesler, M. G. and M. M. Kessler (1958) *World Pet.* **29**, 60.

Kevorkian, A. K. (1975) Structural aspects of large dynamic systems. In *Proceedings of the 6th Triennial World Congress IFAC (IFAC/75)*, pp. 24–30. Boston/Cambridge, Massachusetts.

Kevorkian, A. K. and J. Snoek (1973) Theory and applications in solving large sets of non-linear simultaneous equations. In *Decomposition of Large-Scale Problems* (ed. D. M. Himmelblau), pp. 467–89. North-Holland, Amsterdam.

Kubicek, M., V. Hlavacek and F. Prochaska (1976) Global modular Newton–Raphson technique for simulation of an interconnected plant applied to complex rectification columns. *Chem. Eng. Sci.* **31**, 277–84.

Ledet, W. D. and D. M. Himmelblau (1970) Decomposition procedures for solving large scale systems. *Adv. Chem. Eng.* **8**, 186.

Lee, W. and D. F. Rudd (1966) On the ordering of recycle calculations. *AIChE J.* **12**, 1184–90.

Leigh, M. J., G. D. D. Jackson and R. W. H. Sargent (1974) SPEED-UP – A computer-based system for the design of chemical processes. Paper presented at CAD-74, Imperial College, London, England, Sept. 24–7.

Mah, R. S. H. (1974) Recent development in process design. In *Symposium on Basic Questions of Design Theory*. Columbia University Press, New York.

Marquardt, D. W. (1963) An algorithm for least-squares estimation of non-linear parameters. *SIAM J.* **11**, 431–41.

Motard, R. L. and H. M. Lee (1971) *CHESS, Chemical Engineering User's Guide*, 3rd edn. University of Houston, Texas.

Motard, R. L., M. Shacham and E. M. Rosen (1975) Steady state chemical process simulation. *AIChE J.* **21**, 417–36.

Motard, R. L. and A. W. Westerberg (1978) Exclusive tear sets for flowsheets. To be submitted.

Orbach, O. and C. M. Crowe (1971) Convergence promotion in the simulation of chemical processes with recycle – The dominant eigenvalue method. *Can. J. Chem. Eng.* **49**, 509.

Perry, R. H. and C. H. Chilton (1973) *Chemical Engineering Handbook*, 5th edn, pp. 10–39 – 10–42. McGraw-Hill, New York.

Petlyuk, F. B., V. M. Platonov and D. M. Slavinskii (1965) Thermodynamically optimal method for separating multicomponent mixtures. *Int. Chem. Eng.* **5**, 555–61. Article first published in *Khim. Prom.* **41**, 46–51.

Pho, T. K. and L. Lapidus (1973) Topics in computer aided design. I. An optimum tearing algorithm for recycle systems. *AIChE J.* **19**, 1170.

Prigogine, I. and R. Defay (1954) *Chemical Thermodynamics* (transl. D. H. Everett), pp. 187–8. Longman, London.

Ramirez, W. F. and C. R. Vestal (1972) Algorithms for structuring design calculations. *Chem. Eng. Sci.* **27**, 2243.

Reid, J. K. (1976) Sparse matrices. *AERE Harwell Report CSS-31*. Atomic Energy Research Establishment, Oxfordshire, England.

Reid, R. C., J. M. Prausnitz and T. K. Sherwood (1977) *The Properties of Gases and Liquids*, 3rd edn. McGraw-Hill, New York.

Rosen, E. M. (1962) A machine computation method for performing material balances. *Chem. Eng. Prog.* **58**, 69–73.

Rosen, E. M. and A. C. Pauls (1977) Computer aided chemical process design. *Comput. Chem. Eng.*, **1**, 11.

Rudd, D. F. and C. C. Watson (1968) *Strategy of Process Engineering*, pp. 34–66. John Wiley, New York.

Sargent, R. W. H. (1967) Integrated design and optimization of processes. *Chem. Eng. Prog.* **63**, 71–8.

Sargent, R. W. H. and A. W. Westerberg (1964) 'SPEED-UP' in chemical engineering design. *Trans. Inst. Chem. Eng.* (London) **42**, 190–7.

Seader, J. D., W. D. Seider and A. C. Pauls (1974) *FLOWTRAN Simulation. An Introduction – CACHE Committee*. Ulrich's Books, Ann Arbor, Michigan.

Simon, H. A. (1969) *The Sciences of the Artificial*. MIT Press, Cambridge, Massachusetts.

Smith, G. W. and R. B. Walford (1975) The identification of a minimal feedback vertex set of a directed graph. *IEEE Trans. Circuits Syst.* **CAS-22**, 9–15.

Soylemez, S. and W. D. Seider (1973) A new technique for precedence ordering chemical process equation sets. *AIChE J.* **19**, 934.

Stadtherr, M. A., W. A. Gifford and L. E. Scriven (1974) Efficient solution of sparse set of design equations. *Chem. Eng. Sci.* **29**, 1025.

Steward, D. V. (1962) On an approach to techniques for the analysis of the structure of large systems of equations. *SIAM Rev.* **4**, 321–42.

Steward, D. V. (1965) Partitioning and tearing systems of equations. *SIAM J. Numer. Anal.* **2**, 345–65.

Tarjan, R. (1972) Depth-first search and linear graph algorithms. *SIAM J. Comput.* **1**, 146–60.

Umeda, T. and M. Nishio (1972) Comparison between sequential and simultaneous approaches in process simulation. *Ind. Eng. Chem. Process Des. Dev.* **11**, 153.

Upadhye, R. S. and E. A. Grens (1972) An efficient algorithm for optimum decomposition of recycle systems. *AIChE J.* **18**, 533.

Upadhye, R. S. and E. A. Grens (1975) Selection of decompositions for process simulation. *AIChE J.* **21**, 136.

Wegstein, J. H. (1958) *Commun. Assoc. Comput. Mach.* **1**, 9.

Westerberg, A. W. and C. J. deBrosse (1973) An optimization algorithm for structured design systems. *AIChE J.* **19**, 355.

Westerberg, A. W. and F. C. Edie (1971*a*) Computer aided design. I. Enhancing convergence properties by the choice of output variable assignments in the solution of sparse equation sets. *Chem. Eng. J.* **2**, 9.

Westerberg, A. W. and F. C. Edie (1971*b*) Computer aided design. II. An approach to convergence and tearing in the solution of sparse equation sets. *Chem. Eng. J.* **2**, 17.

Westerberg, A. W. and J. V. Shah (1978) EROS – A program for quick evaluation of alternate energy recovery systems. *Comput. Chem. Eng. J.*, (in press).

Young. D. M. (1971) *Iterative Solution of Large Linear Systems*. Academic Press, New York.

Index

absorber examples
   dilute gas, 199–201
   two plate, 90–101
acceleration methods, 67–77, 136
acentric factor, 104
activity coefficients, 109
adiabatic flash, 111
ALGOL, 205
Amundson, N. R., 226
analysis in design process, 1, 2
Antoine equations, 33, 230
approach temperature, 174, 176
   see also crossover temperature
area-elimination method, 30–1
assignment problem, 79

backsubstitution, 38
Barkley, R. W., 89, 148, 152
basic block diagram, 121–9, 180–8
   compressor, 126
   flash unit, 123
   heat exchanger, 125
   mixer, 121
   pump, 126
   reactor, 124
   simple flowsheet, 128, 181
   stream splitter, 125
   user specifications, 127
   valve, 126
Bayer alumina preparation plant example, 213–17
Bending, M. J., 42, 220
Biess, G., 148
binary interaction parameters, 104, 107
bleed stream, 131
blending, 203
block diagram, *see* functional block diagram *or* basic block diagram
bounded secant method, 30–2, 34–5, 111
Broyden, C. G., 63
Broyden's method, 60, 63–5
bubble point, 110, 111, 170, 172
butane oxidation example, 209–15

CADC, *see* Computer Aided Design Centre

Carnahan, B., 51
cascade module in SYMBOL, 207
cascade of units, 159–60
catalytic cracker, 203, 205
cause-and-effect relationship, 97
Cavett, R. H., 74
Cavett problem, 74–7
check sum, 37
chemical equilibrium, 115–20, 232–8
chemical potential, 109, 111, 114, 119, 232–8
chemical structure, 102, 107, 108
CHESS, viii
Cheung, L. K., 89, 148
Christensen, J. H., 90, 100, 148, 152
code strings, 46
Colebrook equation, 224
column of stages, 229
COMPAID, 8
compressibility, 109
compressor, 132–8, 146
   basic block diagram, 126
   degrees of freedom, 126
   isentropic, 112
computational sequence, 17
computer-aided design, 240
Computer Aided Design Centre (CADC), 7, 131, 207
CONCEPT, viii, 131–40, 146
condensers, 229
connection equations, 14, 20, 23, 161, 177
conservation equations, 195, 218, 220, 226, 233
consistency checking, 103
constitutive equations, 195, 218, 220, 226, 233
constraint mode in SYMBOL, 207–8
constraints, *see* equality constraints *or* inequality constraints
control blocks, 130, 140–7
convergence acceleration (promotion), *see under* sequential modular approach *and under* nonlinear equation solving, multidimensional
convergence efficiency, 237

convergence rate, *see* first-order convergence, superlinear convergence *and* second-order (quadratic) convergence
convergence robustness, 237
convergence tolerance, 136
coolers, 229
cooling curve, 172
costing, 10–13, 17, 18, 21, 24, 138, 173
  of a heat exchanger, 170
critical properties, 104
crossover temperature, 169
  see also approach temperature
Crowe, C. M., 70–2, 213
crystallizers, 207

data, physical property
  generation, 103
  library of raw data, 103, 105
  management, 103
  parameter library, 103, 105, 106, 107
  prediction of missing data, 103, 104
data cycle, 102–5
data fitting, 103, 104
data preparation forms in SYMBOL, 207, 212, 214–15
Davy Computing Ltd, 8
Dean, J. A., 35
deBrosse, C. J., 167, 179
debutanizer example, 229, 230
decision variables, 14, 89–101, 163, 166–7, 176, 178, 179
  see also degrees of freedom *and* tear and/or decision variables
Defay, R., 46
degrees of freedom, 89, 180, 195, 202, 213
  in a flowsheet, 127–9
  in a process stream, 113, 115–20
  in a unit model, 113, 120–7; compressor, 126; extent-of-conversion reactor, 123–4; flash, 122–3; heat exchanger, 125–6; mixer, 120–2; pump, 126; stream splitter, 125; valve, 126
DEM, *see* dominant eigenvalue method
Dennis, J. R., Jr, 65
density, 109, 111
design, *see* process design
design calculations, 23
design specifications, *see* user specifications (constraints)
dew point, 110, 111, 170, 172
diffusivity, 104, 109
directed graph
  for equations, 79–89
  for flowsheet (functional block diagram), 148
direct linearization, *see* linearization
direct substitution, *see* successive (direct) substitution

distillation column, 140, 195, 203, 225–32
dominant eigenvalue method (DEM), 67, 70–2, 76–7
double tearing, 157
Duff, I., 42, 80, 240
Duhem's theorem, 116, 124
dynamics, 5, 239, 240

ECONOMIST, 138
Edie, F. C., 89, 90
efficiency of convergence, 237
engineering line diagram (ELD), 5
enthalpy, 109, 110, 111, 112, 114
entropy, 109, 110, 111, 112
  of formation, 109
environmental considerations, 200, 239
equality constraints, 195, 218, 220, 227, 238
  module in SYMBOL, 208
  see also user specifications
equations of state, 109
equation-solving approaches
  comparision of, 24–6, 191–3
  description of, 23–4, chapters 7, 8, 9
equilibrium, *see* chemical equilibrium, phase equilibrium, pressure equilibrium *or* thermal equilibrium
equilibrium constants, 235
equipment parameters, 15–26, 130, 132–6, 138, 139, 140–7, 162, 172, 181
Evans, L. B., 240
exclusive tear set, 157–8
explicit iteration, 55, 67, 99, 143, 179
extensive variables, 114
extent-of-conversion reactor, 113, 124, 202
extent of reaction, 117, 195
extraction train example, 218–22

feasible point, 179
file of an unused variable, 43
filters, 207
first-order convergence, 49, 228, 237, 239
flash unit, 85–9, 109, 110–11, 131–8
  adiabatic, 111
  basic block diagram, 123
  Cavett problem, 74–7
  degrees of freedom, 113, 122–3
  implicit tears, 144–5
  isothermal, 111
  one-dimensional nonlinear equation example, 32–5
  three-flash-drum example, 196–9
  three-phase, 111
Flexible Flowsheet Program (Kelloggs), 9
flow divider, *see* stream splitter (flow divider)
Flower, J. R., viii, 240

flowsheeting, *see* process flowsheeting
flow splitter, *see* stream splitter (flow divider)
FLOWTRAN, viii
Forder, G. J., 148, 155
FORTRAN, 9, 205, 206
Fox, L., 35
free energy of reaction, 109
fugacity, 109
fugacity coefficients, 109
functional block diagram, 132, 133, 146–7, 148, 149
functional (unit) module, 132, 146
   environment of, 139
   examples in CONCEPT, 132–6
   user-written, 138–40

Gauss elimination, 36–9
   logical, 97–9
Gauss–Jordan method (linear equations), 39, 40
Gauss–Seidel iteration method, 36
generalized secant method, 57–65
   update formulas, 60–5
Gibbs–Duhem theorem, 52, 233
Gibbs free energy, 109, 111, 117
Gifford, W. A., 89
graph, directed, *see* directed graph
Grens, E. A., 148, 152–4, 156, 157, 158
groups
   equations, 85
   units in a flowsheet, 85, 148–50, 188
Guardabassi, G., 89, 148
Gupta, P. K., 79, 82

half-interval method, 30–2
halibut-liver-oil extraction train example, 218–22
Harary, F., 78
Harwell library, 240
heat capacity, 104, 109
heaters, 229
heat exchanger, 111, 112, 169–70, 179, 229
   basic block diagram, 125
   degrees of freedom, 113, 125–6
   partitioning into zones, 170
heat exchanger networks, 165–80
heat of formation, 109
heat of mixing, 109
heat of reaction, 109
heat transfer coefficients, 170, 173
Hendry, J. E., 240
hexene-from-propylene example, 131–8
hierarchical decomposition, 163, 180, 189, 193
Himmelblau, D. M., 78, 89
Hlavacek, V., 24, 240
Householder's formula, 64

Hutchison, H. P., 24, 42, 51, 52, 148, 155, 207, 220, 226–231
hydrocracker unit, 205

ICI (Imperial Chemical Industries Ltd), 8
IHI (Ishikawajimi-Harima Heavy Industries Co. Ltd), 7
implicit block module, 143–7, 148
implicit iteration, 55, 99, 143, 179
incidence matrix
   for equations/variables, 78–101, 177–8;
      definition, 78, 85–6
   for flowsheets using equation-solving approach, 181, 183, 189
   loop/stream for units, 156–60
independent equations, 113
independent reactions, 117, 162, 207, 232, 234
indirect methods for solving nonlinear equations, 54–101
inequality constraints, 166–7, 169, 171, 173, 174, 175, 192
   conversion to equality constraints, 176
information flow, 181, 209
information flow diagram, 80, 162–5
   for SYMBOL, 209
information streams, *see* streams, nonstandard (information)
intensive variables, 114
inverse matrix, 39–42, 47, 67
IPAC, 7
isentropic compressor, 112
ISOPEDAC, 8
isothermal flash, *see under* flash unit

Jackson, G. D. D., viii, 24, 161
Jacobi iteration method, 36, 68, 69
Jacobian matrix, 51
Janicke, W., 148
JOE (simultaneous overrelaxation), 68–70, 76–7
Johns, W., 148

Kehat, E., 240
Kesler, M. G., 9
Kessler, M. M., 9
Kevorkian, A. K., 89
Key, J. E., 42
Kirchoff's law, 196
Kremser equations, 195
Kronecker delta, 228
Kubicek, M., 24
Kuh, E. S., 89, 148

laminar flow, 223
Lapidus, L., 89, 148, 152
least squares, 101, 104
   Marquardt method, 53–4
Ledet, W. D., 89

Lee, H. M., viii
Lee, W., 148
Leigh, M. J., viii, 24, 161
linear equations, 3, 18–23, 25, 49–150,
    chapters 8, 9
  see linear equation solving
linear equation solving, 27
    Gauss elimination, 36–9
    inverse matrix, 39–41
    L/U decomposition, 41–2
    multiple right-hand sides, 39–42
    over and under specification, 44
    singular coefficient matrix, 45
    sparse equations, 42–8
linear programming, 79
linearization, 24, 27, 48–54, 83, 188, 218
logical Gaussian elimination, 97–9
'loopfinder' algorithm, 155
lower bounds, 173
L/U decomposition, 41–2
Luther, H. A., 51

Mah, R. S. H., 240
management example, 200–6
Marquardt, D. W., 53
Marquardt method, 53–4
mathematical service routines, 10–12
matrix inversion, 39–42, 47, 67
mechanical design, 8
methane combustion example, 234–7
missing-data prediction, 103, 104, 108
mixer, 114–15, 171, 194, 196, 229
    basic block diagram, 121
    degrees of freedom, 113, 120–2
    module in SYMBOL, 207
mode, constraint in SYMBOL, 207
model equations, 14, 16, 20, 23
More, J. J., 65
Motard, R. L., viii, 89, 148, 152, 153–60,
    240

naptha separation plant example, 231–2
Newton–Raphson (NR) iterative pro-
    cedure, 49, 51, 72, 76, 225, 226, 237
Newton's method, 28–9
Nishio, M., 18
nonlinear equation solving, multidimen-
    sional, 13, 23, 27, 130, 143, 218
    direct linearization methods, 48–54, 83;
        alternative linearization schemes,
        48–50; Newton–Raphson iterative
        procedure, 51; quasi-linear iterative
        procedure, 51, chapter 9
    indirect methods, 54–101; acceleration
        methods, 65–77, 136; Broyden's
        method, 63–5; generalized secant,
        57–65; sparse nonlinear equations,
        77–101; use of acceleration, 151
    redundant equation sets, 101

nonlinear equation solving, one-dimensional
    half-interval method, 30–1
    Newton's method, 29
    secant method, 29–30
nonsquare nonlinear equation sets, 89–101
nonstandard streams, 140–7
NR, *see* Newton–Raphson iterative proce-
    dure
occurrence matrix, *see* incidence matrix
oil refinery example, 203–6
operability, 5
operating constraints, *see* user specifica-
    tions
operator lists, 46–8
optimization, 1, 2, 90, 131, 166–7
Orbach, O., 70–2
orthogonal vectors, 62–5
outer product, 61
output (destination) module, 132
output variable, 79
overrelaxation, 67–70
oxidation-of-butane example, 209–15

PACER, viii
parameter optimization, 1, 2
  see also optimization
partial solution of linear equations, 43–5
partitioning
    flowsheets, 135, 148–150, 165
    full, 79
    nonsquare equation sets, 91, 100, 187–8
    square equation sets, 77–84, 86, 87, 198
path tracing
    equations, 80
    flowsheets, 148–50, 155
Pauls, A. C., viii, 74–6
PDMS (Pipework Design Management
    System), 7
performance (rating) calculations, 23, 200
performance records, 202
Perry, R. H., 170
Petlyuk, F. B., 229
Petlyuk towers example, 229–31
petroleum boiling fractions, 107
PFD (process flow diagram), 5–6
phase equilibrium, 115–20
phase rule, 113–14
Pho, T. K., 89, 148, 152
physical properties, 3–5, 9–12, 17, 21, 33,
    35, 91, chapter 4, 138, 173
    computerized service, 105–9
    data cycle, 102–5
    equations for equation-solving approach,
        161, 171
    estimation of, 102, 103
    parameter library, 103, 135
    provided in typical service, 109–12
physical property parameter library, 103,
    105, 106, 107, 108, 135

PID (piping and instrumentation diagram),
  5–7
pipe network example, 220–6
Pipework Design Management System
  (PDMS), 7
pipework isometrics, 7
piping design, 7, 8
piping and instrumentation diagram (PID),
  5–7
pivot element, 36
pivot equation, 36, 37
pivot sequence, 83, 101
plant performance records, 202
point data generation, 106
postprocessing unit, 141
Prausnitz, J. M., 112
precedence ordering
  flowsheets, 148–50
  square equation sets, 77–84, 86, 87
preparation forms in SYMBOL, 207, 212,
  214–15
preprocessing unit, 141
pressure drop in pipe, 223
pressure equilibrium, 114, 119
Prigogine, I., 116
process conception, 5
process design
  scenario, 3–4
  the total project, 5–8
process flow diagram (PFD), 5–6
process flowsheeting
  advantages/disadvantages, 4
  alternative approaches, 13–26
  definition, vii, 1–5
  history on computer, 9
  motivation for development, 10–11
process synthesis, 1, 2, 240
Prochaska, F., 24
project engineering, 5, 6
propylene-to-hexene example, 131–8
pseudo raw data, 108
pump, 220
  basic block diagram 126
  degrees of freedom, 126
pump-around circuit, 230
PVT-relationships, 110

QL, *see* quasi-linear iterative procedure
quadratic convergence, *see* second-order
  convergence
quasi-linear (QL) iterative procedure, 51,
  217, chapter 9
quasi-linear process simulator, 238–9

Ramirez, W. F., 90
rank of an unused equation, 43
'rank 1' update formulas, 60–5
Raoult's law, 33

rating (performance) calculations, 23, 200
raw data library, 103, 105, 106, 108
reaction equilibrium (chemical equilibrium), 115–20, 232–8
reactor, 131–8, 162, 211
  basic block diagram, 124
  degrees of freedom, 113, 123–4
  extent-of-conversion, 113, 124, 202
  module in SYMBOL, 207
reboilers, 229
redundant equations, 44, 101, 115, 197
redundant/nonredundant decompositions,
  152–60
refinery example, 203–6
Reid, J. K., 80, 240
Reid, R. C., 112
relative volatility, 227, 230
reliability, 8
replacement rule, 152–4, 158
retention factors for halibut liver oil, 219
roadmapping problem, 104
robustness of convergence, 237
Rosen, E. M., 18, 74–6, 240
Rudd, D. F., 148, 152, 240

SACDA, 206
safety, 8
Sargent, R. W. H., viii, 24, 80, 148, 161,
  240
scaling, 36
scenario for flowsheeting, 3–4
Scriven, L. E., 89
Seader, J. D., viii, 240
secant method (SM)
  generalized (multidimensional), 57–65,
    191
  one equation in one unknown, 29, 56,
    111, 143
  Wegstein's method, 65–7, 70, 136
second-order (quadratic) convergence, 49,
  50, 51, 65, 228, 237, 239
segments, stream, 167, 173
Seider, W. D., viii, 89
separation trains, 229–31
separator module in SYMBOL, 208
sequential modular approach, 54, 85, 163,
  180, 192
  acceleration, 151
  acceleration methods for, 67–77; com-
    parison of acceleration methods,
    76–7; dominant eigenvalue method,
    70–2; overrelaxation, 67–70; secant
    (or Wegstein's) method, 72–6
  comparison with other approaches,
    24–6, 130
  control blocks and nonstandard streams,
    140–7
  description of, 15–18, chapter 6
  example use of, 130–8

sequential modular approach—*contd.*
  partitioning and precedence ordering, 148–50
  tearing, 150–60
serial number, 46
Shacham, M., 240
Shah, J. V., 24, 100, 165–180
Shannon, P. T., 162
Sherwood, T. K., 112
Shewchuk, C. F., 24, 51, 52, 226–231
sidestream coolers, 229
sidestreams, 230
Simon, H. A., 193
simulation by quasi-linear methods, 238–9
simulation model, 132
simultaneous modular approach
  comparison with other approaches, 24–6
  description of, 18–23
simultaneous overrelaxation (JOE), 68–70, 76–7
singular coefficient matrix, 45, 101
  structural, 79, 83
slack variable, 176, 178, 179
SM, *see* secant method, Wegstein's method
Smith, G. W., 89, 148
Snock, J., 89
Solemez, S., 89
source module
  in CONCEPT, 132
  in SYMBOL, 208
spares, 5
sparse equations
  definition, 27
  linear equations, 42–8, 195, 200, 225, 231, 238, 239
  literature references, 240
  nonlinear equations, 77–101; nonsquare equation sets, 89–101; square equation sets, 77–89
specifications, *see* user specifications (constraints)
SPEED-UP, viii, 161
Sprague, C. R., 240
square nonlinear equations, 77–89
Stadherr, M. A., 89
statistical fitting
  data fitting, 103–4
  least squares, 101, 104
steepest descent, 53, 54
Steward, D. G., 240
Steward, D. V., 79, 81, 89
Steward, path, 81, 82, 84
stoichiometric coefficients, 116, 134, 195, 232
stream junctions, 229
stream loop, 155–160
stream segments, 167, 173
stream splitter (flow divider), 170, 180, 194, 229

basic block diagram, 125
degrees of freedom, 113, 125
  see also bleed stream
stream string, 153
streams, nonstandard (information), 140–7
string of streams, 153
string of units, 148–50
  full, 153
structural optimization, 1, 2
structural singularity, 79, 83
structure of equations, *see* sparse equations
successive (direct) substitution, 55, 56, 153
  acceleration methods for, 67–77
superlinear convergence, 65
surface tension, 109, 111
SYMBOL, viii, 195, 207–17
SYMMAN, 206
synthesis, 1, 2, 240

Tarjan, R., 80
tear and/or decision variables, 92–101, 180–8
tear equations, 88, 184–8, 190–1
tearing
  equations, 24, 26, 27, 85, 86–9, 90–101, 179
  flowsheet, 17–18, 85, 135, 148–60; various criteria for, 152
  flowsheet using equation-solving approach, chapter 7
  implicit/explicit, 143–7
  Upadhye and Grens algorithm, 152–3
tear set multiplicity, 158–60
tear streams, 135, 143
tear variables, 85, 87, 143, 179, 190–1
  see tear and/or decision variables
temperature setters, 229
thermal conductivity, 104, 109
thermal equilibrium, 114, 119
thermal expansion coefficient, 109
thermodynamic properties, 110
three-phase flash, 111
transient analysis (dynamics), 5, 239, 240
transport properties, 110
triangularization, 37
turbulent flow, 223

ULD (utility line diagram), 7
Umeda, T., 18
unit loops, 153–160
unit modules, *see* functional (unit) module
unit strings, 148–50, 153
Upadhye, R. S., 148, 152–3, 156, 157, 158
upper bounds, 173
user specifications (constraints), 174, 181, 192, 194, 202, 205, 206, 213, 220, 238
utility line diagram (ULD), 7

validity checking, 103
valve
  basic block diagram, 126
  degrees of freedom, 126
vapour–liquid equilibrium, (VLE), 109, 110–11, 226–8
  see also flash unit
Vestal, C. R., 90
viscosity, 104, 109
VLE, *see* vapour–liquid equilibrium

Walford, R. B., 89, 148
Wegstein, J. H., 65, 136
Wegstein's method, *see under* secant method
Westerberg, A. W., 24, 80, 89, 90, 100, 148, 153–60, 165–80, 167, 179
Whitehead, B. D., viii, 240
Wilkes, J. O., 51

Young, D. M., 68, 69, 70

Made in the USA
Monee, IL
07 July 2026